ROYAL COLLEGE OF ART SCHOOLS TECHNOLOGY PROJECT

ADVANCED MANUFACTURING DESIGN & TECHNOLOGY

TEACHER'S GUIDE

POST-16

Hodder & Stoughton
A MEMBER OF THE HODDER HEADLINE GROUP

The cover illustration was prepared by fab 4 studio

A catalogue record for this title is available from The British Library

ISBN 0 340 705299
First published 1999

Impression number	10	9	8	7	6	5	4	3	2	1
Year	2003		2002		2001		2000		1999	

Typeset by Wearset, Boldon, Tyne and Wear.
Printed in Great Britain for Hodder & Stoughton Educational, a division of Hodder Headline Plc, 338 Euston Road, London NW1 3BH by Hobbs the Printers, Totton, Hants.

CONTENTS

CHAPTER 1

INTRODUCTION

The **Royal College of Art Schools Technology Project** (RCA STP) publishes material to support a cohesive and progressive course in Design and Technology throughout the secondary age range, 11–19. The complete course consists of:

1. **KS3 D&T Challenges books.** These comprise a pair of books for students and teachers for each year: Y7, Y8 and Y9, coded red, green and blue respectively. A **Course Guide** for KS3 is also available to teachers and is a valuable resource to support the principles of Design and Technology education.

2. **KS4 D&T Routes books.** There are six routes books for students:

- The **Core Book**, which is designed to guide them on their personal route through the 14–16 phase and help them work independently of their teacher as much as possible. It is also suitable for teachers to use with classes, groups of students or individuals in the traditional text book manner.
- Five **Focus Area books** covering *Resistant Materials*, *Control Products*, *Food*, *Textiles* and *Graphic Products* expand on the general principles in the *D&T Routes Core Book* for students as specific to the different materials and learning demands.

The **D&T Routes Teacher's Resource for KS4** provides routes for teachers through the maze of requirements and opportunities that face them in structuring Design and Technology experiences for students.

3. **Advanced courses**

Advanced Manufacturing, Design & Technology student book supported by this Guide.

The features of this course of publications

These books provide:

- a major resource for students aged 16–19, extending their understanding to encompass a wide range of industrial contexts
- a complete response to the requirements of the National Curriculum order for Key Stages 3 and 4. All aspects of the requirements of the Orders are provided for
- easy accessibility – they can be adopted and used directly or adapted as necessary to suit different schools and colleges
- a clear structure for teaching and learning, distinguishing between teacher-directed learning and student responsibilities including a great deal of support for increasing student self-reliance
- continuity: steady, continuous support for students building progression though KS4 and post-16 from foundations laid at Y7/8/9
- teachers' management guidance for whole key stages, and broken down into year by year approaches for KS3. This is coherent, from a course-planning level down to detailed advice for individual tasks.
- assessment guidance compatible with both National Curriculum and examination agency specifications for the new millennium
- frequent industrial insights: expectations of students are repeatedly related to what goes on in the adult world, including in-depth case studies used to support learning

- sections devoted to developing teachers' understanding of design and technology and manufacturing.

The Project's principles

The Project has a number of basic principles:

- demonstrating the relevance to the adult world of design and technology both today and in the future
- acknowledging and developing from a prior learning base and prior experiences
- attention to the needs of individual students, challenging and supporting all
- progression in terms of students' capability, responsibility, autonomy, awareness and pride in their own learning, knowledge and understanding, awareness of others' needs, understanding of the ways that design and technology affects other people, creativity, sensitivity, self-confidence and self-reliance
- making a technologically complex mode of living approachable
- making meaningful use of matter studied in other areas of the curriculum
- drawing out the relationship between the nature of designing and learning
- encouraging teachers and others to work together.

What are these prior experiences?

Students coming from KS4 courses will have to have covered the requirements of the National Curriculum Order. For most students this will be a GCSE course in Design and Technology, although some will have followed a GNVQ course in Manufacturing or possibly Engineering.

The basic requirements of the KS4 National Curriculum are that students should develop their capability through **designing and making assignments**.

These should be supported by:

- focused practical tasks to develop skills and understanding
- product evaluation
- activities that relate their designing and making to the uses of systems and control and industrial practices.

However, GCSE students may have specialised in a focus area such as Resistant Materials, Food, Textiles, Control or Graphic Products. Some students may have worked in more than one of these materials areas. GNVQ manufacturing students will have experience of all manufacturing sectors but will have specialised in two.

Independent learners

A particular aim of the RCA STP is to achieve the greatest possible independence in students.

It is a common expectation of GCSE and GNVQ at this level that students increasingly determine their own direction of work. All GCSE D&T syllabuses require a larger scale project which students are expected to direct for themselves. However, it is recognised that many teachers will set designing and making assignments, particularly in Y10, to ensure that students adequately cover the expectations of the syllabus or course units.

For students on Advanced courses there is an even greater expectation that they will take much more responsibility for planning, managing, organising and evaluating their own work, and this principle of the RCA Project is developed further in these Manufacturing, Design & Technology books.

Case studies

A great many case studies are provided in the Project's books to give students insights into the adult worlds of designing and making. They are useful in both guiding students in their project work and also in increasing their knowledge and understanding. This is particularly true in extending understanding of industrial manufacturing approaches in contrast to students' experience of one-off craft making. This feature is extended in the Advanced books.

High quality learning

Everything a school does is ultimately concerned with the quality of each individual's learning. Similarly, every aspect of a teacher's job is ultimately concerned with the same thing. This book is designed to help teachers of Design, Technology and Manufacturing courses in their task to attain higher standards of learning for their students. Some parts are focused closely on the day-by-day activities in classrooms while others are broader, relating to the underpinning which any teacher, department or school needs to have in place to support educational progress.

The RCA Project team will be happy if every reader constantly queries the value of this book in terms of improving learning. Nothing we suggest should be so remote from the experiences of teachers that it can not be applied in schools of all sorts to improve the quality of learning outcomes.

Drawing on a simple but powerful idea which stems from industry, people in education are recognising that to be never satisfied is a healthy state of mind and that their professional concern should be to seek **continuous improvement**.

It is clear from the work of the school improvement movement that achieving this requires many factors to be working together, but while system-wide or school-wide concerns have their proper place, this book focuses on the Design and Technology teacher, the Design and Technology department and the relationship between these and the more immediate aspects which impinge on them in their quest to continuously improve quality: students, resources, colleagues and senior managers.

Summary: What does this mean for the RCA Project's Advanced books?

Applying these principles to the RCA Project's Advanced books means:

- building on the firm foundations and principles established by the *D&T Challenges* (KS3) and the *D&T Routes* (KS4) series of books for students and teachers
- an increasing need to develop students' **autonomy and responsibility** for planning, organising, managing and evaluating their own work and their own learning
- a need to develop sustainable partnerships involving students, teachers and people from business and industry to the **mutual benefit** of all the individuals concerned and their institutions/organisations

- establishing and applying a philosophy of **continuous improvement** throughout all aspects of the work of individuals and their institutions/organisations

This means that the Project focuses on the following in its Advanced materials:

- Supporting learning rather than teaching. Although parts of this Guide give advice and guidance to teachers and mentors to help them support students, it is recognised that these teachers and mentors are also learners and need to be similarly supported in their learning and professional development.
- Advanced studies as part of and preparation for lifelong learning.
- The incorporation of key skills and the skills that make people employable.
- Manufacturing and designing for manufacturing with frequent insights into the world of professional designers and companies.

Using this Guide

This Guide contains information and guidance on:

- the features of good students' work on Advanced courses in Manufacturing, Design and Technology
- teaching and learning issues
- continuity and progression
- differentiation in Advanced courses
- promoting increased student autonomy
- planning an Advanced course
- developing good assignments
- a range of management issues
- teachers and others working together to support students' learning – sustainable partnerships
- advice on making effective use of the student book.

This Guide should be used alongside the companion *Advanced Manufacturing, Design & Technology* student book; much of that material is equally appropriate for teachers and is not repeated in this Guide although references are made where appropriate. You should also refer to the *D&T Routes Teacher's Resource for KS4* which establishes the principles for much

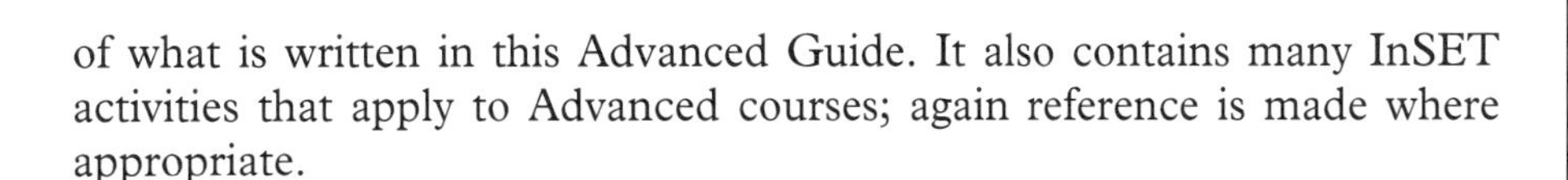

of what is written in this Advanced Guide. It also contains many InSET activities that apply to Advanced courses; again reference is made where appropriate.

Continuous improvement

One of the key features of the Student book is the use of a continuous improvement theme throughout. The intention is to clearly establish in students' minds the basic principle that there is often a better way of doing things, that it can always be done better next time and that they should always be looking to improve the quality of their work. The concept of continuous improvement is the key to Total Quality Management (TQM) in industrial, commercial and other organisational contexts. It has equally important applications in education and in learning. The idea is also developed in the *D&T Routes Teacher's Resource for KS4*.

The basic principles of continuous improvement are:

- changing attitudes and practices – questioning existing procedures to improve their effectiveness and moving from 'there is no alternative' to 'there must be a better way'.
- working systematically and asking questions such as
  - why are we doing this?
  - why are we doing it this way?
  - what are we really trying to achieve?
  - why are we aiming for those particular outcomes?
  - how can we achieve them?
  - what alternative ways are there?
  - how will we know when we have got there?
- making full use of the skills and abilities of everyone in the team.

To reinforce this idea and help you apply it, each section of the book contains InSET activities designed to help you use the information to improve the quality of your Advanced courses. The activities are designed to develop a better understanding of the course approach and content, and also to add structure to planning and implementing the course. These will help you step-by-step through some difficult teaching and learning issues. They are valuable for professional development and preparing for inspections. Choosing relevant activities and completing them as a faculty or department will also enhance the strength of your team and develop a shared understanding of your subject. Use them at your routine department or faculty meetings or on InSET days. Remember, your team may well include people not on the school or college teaching staff; they should also take part in these activities as far as possible.

There should be stated objectives for your InSET activities. They should be based on a clear identification of institutional needs (relating to the development plan) and should start from teachers' current levels of knowledge and skills. Development plans are now an important part of overall planning strategies for schools. From the process involved in drawing up such plans, staff can learn much about areas of strength and weakness and be clearer about the InSET required.

The process to follow:

- Review the activities in this book.
- Consider those which are relevant to your faculty or department or teaching team.
- What are your priorities? Decide these as a team.
- Share the purposes of any task across the team.

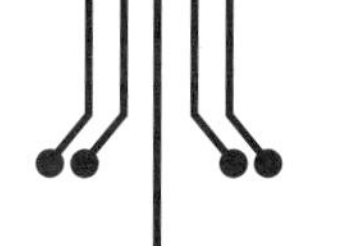

- Should it be done alone, as a team, or with some individual preparation time?
- Ensure there is time to complete the activity – avoid interruptions.
- Agree the ground rules for group behaviour.
- Avoid making individuals feel inadequate or 'got at'.
- Set clear time limits and be clear about what outcome you intend.
- Give space to people at the end to say how they felt.
- How will it be recorded and fed back? To whom?

Points for heads of departments or faculties and team leaders

Work from where your staff are now, and celebrate the successes and strengths of existing work and approaches. Help each staff member to create a personal vision and a development plan for their own contributions.

Respect the worth of every individual and help to build confidence in them. Contributions from individual staff should be based on current strengths rather than exposing weaknesses.

Create time for development that is planned and flexible enough to respond to the needs of the department or individuals. Although time–consuming, everyone will benefit and feel better about their role.

Sensitive behaviour to colleagues will build a positive ethos and dynamic culture in Design and Technology. Encouraging lively debate in order to reflect the ever changing nature of educational challenges will help you all enjoy a positive, collaborative atmosphere that will do much to protect you from the negative pressures that are often brought to bear on education from outside.

You will also find a number of **Continuous Professional Development (CPD) activities** within the book. These are designed to help you reflect on your own practice and apply the principles of continuous improvement to your own professional development.

Who should use this Guide?

This book is designed to be used by everyone involved in supporting students on Advanced courses in Manufacturing, Design and Technology. This includes teachers, industrial mentors, partners in industry, colleges, Universities and in the community, and any adults who come into schools to work with students, such as Neighbourhood Engineers. We have used the phrase **industrial mentors** as a form of shorthand to include all of these people.

The Guide should be used in a collaborative and complementary way; it is a shared resource and is not intended to be the property of any one individual or group. The InSET activities included in the Guide are not just for use by the teaching staff in the school/college.

InSET – Building a support team

It is invaluable to build a team of people to support the students. This team needs a shared understanding of the principles behind the course and the way it will be delivered. In this way it becomes possible for different members of the team to clearly identify the contributions they can make and how these fit into the 'bigger picture'.

The ideal way to achieve this is to have team building sessions before the course begins and the following activity is designed to provide a structure and ideas for such a session. This could involve the students as well as the others helping with the delivery of the course. However, this may not always be possible and other ways need to be found to achieve the same outcomes. Some suggestions for this are also included.

Team building activity

The outcomes of this activity should be:

- a shared understanding of the principles behind the course and of the course requirements
- a sense of common purpose
- a clear understanding of the range of student activities that will be used on the course and their purpose
- each individual being clear about their contribution, what they need to do and when and how it fits into the delivery of the complete course
- an outline plan of the course with the activities and deadlines mapped out.

1. Go through the requirements of the course and ask each person to note down what this means to them.
2. Go around and collect these ideas.
3. Discuss to reach a common understanding.
4. Ask each person what they could contribute to meeting these requirements.
5. Ask each person what they hope to get out of the experience.
6. Use this to plan individual and group contributions.

Other suggestions for team building and supporting the course delivery:

- produce a course handbook containing all relevant information including the course plan
- hold regular planning, review and evaluation meetings with students, industrial mentors and others
- pair members of the teaching staff with industrial mentors
- pair students with industrial mentors.

CHAPTER 2 ADVANCED LEVEL COURSES IN MANUFACTURING, DESIGN AND TECHNOLOGY

Why Advanced level manufacturing, design and technology courses?

For many years debates about change in the post-16 curriculum have emphasised the need for all students to have a broader education in the 16–19 period. The need for courses to cater for varied learning and assessment styles has been recognised with the advent of the GNVQ, and the revised GCE A and A/S level specifications from 2000. As these unfolded, the overriding change that has dominated the 16–19 sector has been increased participation rates. This only brings England into closer alignment with other developed countries and can be expected to continue.

Changes in levels of participation and the broadening of the range of subjects studied combine to increase the number and range of students involved in manufacturing, design and technology courses and these two books seek to support schools and colleges facing these concerns. There has been clear evidence in many schools that, especially if Intermediate GNVQ courses are not provided, GCE A and A/S level classes have housed students who do not learn effectively in that context. Typically, fairly average achievers have dominated and they may or may not have found GCE A level courses suitable. Broadening the range of students means in particular that more academically successful ones are joining manufacturing, design and technology courses, especially at A/S level, increasing the demand on teachers for versatility.

Catering for more academic students

The Dearing review of the National Curriculum identified a '*very large number of students whose interests and talents lie primarily in academic study*'. Traditionally, post-16, this would mean those who follow a sixth form education, those in grammar schools and those taking GCE A levels, going on to university. Many teachers' definitions of academic would preclude Design and Technology – they would not see it as an academic 'subject'. In practice, the antecedent subjects to Design and Technology (CDT, Home Economics etc.) usually had very small A level groups which were, in A level terms, a skewed cohort. That is, they were often not clearly suited to A level learning approaches, or A level learning approaches were not suited to them. A few departments captured a small number of strong A level students (i.e. those following a three or four A level course, tracked for a university place in a mainstream subject). Rarely though, were these the highest flyers.

Manufacturing, design and technology as advanced general education

The history of the design and technology area of the curriculum does not suggest that 'high flyers' will come forward in large numbers for Advanced courses of any type. However, there is increasing support from influential people for a more balanced valuing of what different subjects have to offer, and for the most educated elite from our schools to have a better understanding of the world of commerce and industry, even though they may be destined for careers in law, medicine, the civil or diplomatic services. This latter group will include many of the major decision takers in our future society. For that reason alone, those committed to the place of manufacturing, design and technology in education will want them to continue to be involved in their post-16 education.

There have been some interesting studies which have attempted to re-define popular concepts of intelligence (Sternberg 1985, Gardner, 1984) and broaden the range of characteristics regarded as evidence. Anyone who has attended seminars at Oxford and the Royal College of Art will have been able to identify some similarities and some differences. Probably the differences are less marked the higher up the spectrum of ability you go. High levels of intense concentration, a relentless pursuit of ideas through to an acceptable conclusion, long term diligent commitment to a project once it captures their imagination, a combative enjoyment in taking a position, and a ruthless determination to destroy insubstantial work of any sort are characteristics to be found in able students in any context. But this risks stereotyping, and we must remember others such as those who children now call 'nerds', introvertedly intense (usually) in their concentration on a problem of high particularity; or the blazing extrovert with little apparent commitment to study, who then reveals masterly ability at a stroke; and yet others.

What any of these might react against in an academic approach to D&T is a teacher who lacks depth of understanding, high levels of competence in their subject or a broad awareness of contemporary issues. They may resent more than others any association with the second rate (lessons in a nineteenth century factory-like workshop, a fussy over-concern with trivial details obstructing a grand vision), any over-regulation that lacks intellectual justification (the examiner expects you to do it this way . . .) or any form of organisation or regulation that gets in the way of their progress.

What then, in manufacturing, design and technology, should be stressed for the academically successful? There are implications in much of the above which might guide us in serving the needs of these students. Our teaching and learning approaches need to be flexible and responsive, allowing the pursuit of long term and very personal objectives. Unwavering commitment to a project in hand may have to be tolerated, at the expense of not working systematically through the course specification. Perhaps intense specialisation to a very narrow area may support more general understanding of the subject, and not deny it as we would normally expect.

The relationship between teacher and taught has to be defined on a very equal basis. This remains a threat to those who are insecure in their authority, though no problem to those teachers who are really committed to design-based approaches to learning. Indeed, the latter group know that design-based learning requires such a relationship.

The subject matter must be challenging and be treated in a demanding way. Boundaries also need to be broken. Neither the knowledge base, nor the resources available can be confined to those conveniently available in the classroom or even the school. Bright ideas must be followed through with the least possible obstruction. The thrill of entrepreneurship must be given reign such that the students are pursuing what they need wherever they may find it.

Is this especially for the *academic*? Those familiar with the development of

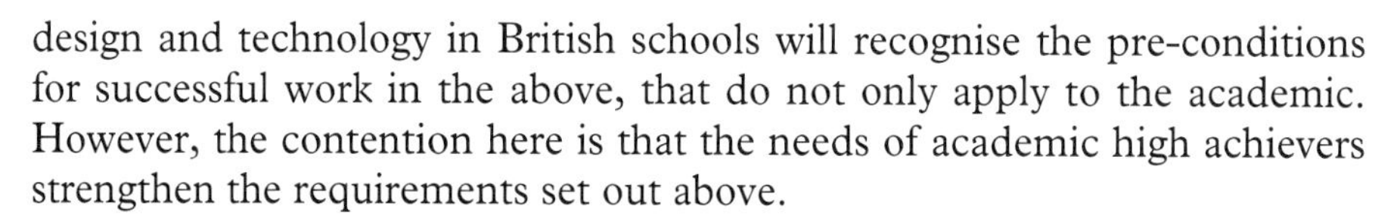

design and technology in British schools will recognise the pre-conditions for successful work in the above, that do not only apply to the academic. However, the contention here is that the needs of academic high achievers strengthen the requirements set out above.

Manufacturing, design and technology courses are often very individualised in their approach, and by nature they require a very diverse range of learning styles, including the analysis and written expression that is at the heart of the humanities, and logic and numerical ratification from the sciences, as well as designing, making and evaluating. Differentiation through varied emphases must therefore result from individual projects, each with their own particular requirements. A poster design does not require the same disciplines as typography any more than a food product requires the same knowledge base as control technology.

Some facets of D&T though, demand different treatment for these students. If we are to exploit the characteristically argumentative tendency they have, then time and space must be given to it. Every teacher knows that many students do not respond well to higher education-like seminar discussions of issues. Yet others thrive on it and they should be accommodated. Similarly, intense interest in the technical, social or environmental issues, visual acuity and understanding of the material culture, are all possibilities for an academic approach that suits some students better than others.

How might this be done? The picture painted here makes demands on the teachers and also on the nature of examination specifications, the school timetable and the working practices in the classroom, studio and workshop. Most particularly, schools do not usually have the student numbers or the flexibility to run alternative courses side-by-side in the manufacturing, design and technology area. Answers therefore must lie with co-teaching for some of the larger class sizes, in flexible assessment schemes to reward different individualised responses and differentiation in teaching styles. Schools and colleges might then be able to secure some separate time for different course groups though this is demanding on staffing. There are many common elements in Advanced courses such as Art & Design, Food Technology, Home Economics, Manufacturing, Design and Technology, Health and Social Care, Engineering, and Business Studies. Schools and colleges offering more than one of these can achieve economies in staffing by combining large groups for some elements and then select which aspects to maintain small, separate groups for, down to individual tutorials.

Learning and teaching styles

The doing of design and technology, i.e. the students experiencing the processes of evaluating designing and making, is the very stuff of the subject. Without this first hand, practical, 'hands-on' experience one can not be said to be studying D&T. However, with this comes a body of knowledge and understanding that enriches and informs these core activities, some of which fall under these headings:

- analysing products and their applications
- understanding product semantics
- design history
- the history of technology
- the impacts of technology on society and the environment
- the nature of designing
- manufacturing methods.

A wider range of teaching styles than usually experienced in manufacturing,

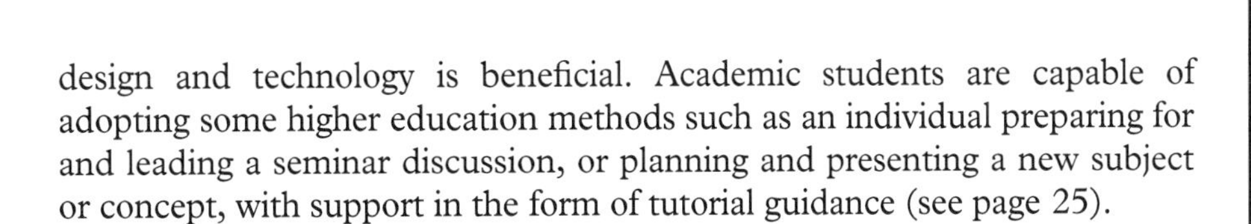

design and technology is beneficial. Academic students are capable of adopting some higher education methods such as an individual preparing for and leading a seminar discussion, or planning and presenting a new subject or concept, with support in the form of tutorial guidance (see page 25).

Here is an example in the manufacturing field:

1 The tutor introduces the four categories of manufacturing to all students.
2 Students in pairs research one category, prepare and give graphic presentations to the rest of the group – the tutor has one advisory session with each pair.
3 Project activity (later?) focuses on design for volume production. The group designs the product, the necessary production system and aids (jigs and fixtures), and part of the group manufactures a batch.
4 The tutor conducts a seminar focusing on the impacts of high volume production on society, the other part of the group are set an essay assignment as a follow up.

This sort of integration and yet differentiation of different learning styles will need to be explored further, both in the final Key Stage of the National Curriculum and post-16.

Further implications for the teachers

In taking this academic approach into the 16–19 age range, a wider range of teaching styles will also need to be developed. Of course, help for this is widely available and at hand. Colleagues teaching traditional A levels will be practised in techniques such as neutral chairing of discussion groups, and tutoring rather than teaching. The students themselves will be strong in skills used and developed in other subjects, such as study skills for information research and the writing of essays. Confidence may be demanded of the teacher, but we are better able to support each other in teaching methods now than teachers were twenty years ago.

Conclusions

Three special characteristics are proposed then for a specifically academic 'spin' to be included in manufacturing, design and technology courses:

- appropriate learning, and therefore teaching, styles
- a different bias to (and scale of) the knowledge and understanding base
- increased flexibility in assessment across GNVQ, A and A/S levels.

Features of manufacturing, design and technology courses

There are a wide range of courses that fit under the umbrella title of manufacturing, design and technology. Some have an emphasis on *product design* and others on *technology*, there are various *focuses*, and there are courses in Art and Design, Food Technology and Engineering. Many of these courses allow students to work in a range of different materials or in one material only, such as food. All Advanced courses in this area share some common features, in particular they all provide opportunities for students to:

- design and make artefacts
- learn about industrial design and manufacturing
- increase technological competence
- understand the many processes through which products are created
- acquire knowledge, skills and attitudes that equip them for careers

- apply knowledge, understanding and skills from many other areas
- take a large degree of personal responsibility for their work.

Manufacturing, design and technology courses require that students bring together knowledge, understanding and skills to create new products or systems, as developing students' 'capability' is the centre of all these courses. Design and technology is an intellectual activity, in that it engages the mind in a process of reasoning. The fact that the reasoning may not be verbal and the outcome is a practical product which has to be shown to satisfy objectives in no way diminishes the intellectual content, and imposes a form of rigour which is not present in solely theoretical reasoning.

The process of designing and manufacturing relies heavily on evaluative judgements and can be seen as:

- an integrated activity undertaken for a design purpose, to meet a perceived need
- requiring judgements to be made about the design process itself as well as about the design proposals
- requiring judgements to be made on sound evidence – sufficient lines of possible development explored to meet the requirements as fully as possible, sufficient breadth and depth of knowledge brought to bear, ideas fully evaluated in relation to the requirements
- an activity that requires a body of knowledge for the purpose of making secure judgements, not as an end in itself.

Manufacturing

Manufacturing industries are the engines of economic growth and these industries are also the most prolific generators and disseminators of new technology. Manufacturing integrates more numerous and varied inputs of goods and services and cultivates a greater variety of skills than many other kinds of activity. For these reasons, an emphasis on manufacturing is essential for any Advanced course in the design and technology area.

Manufacturing as an activity can be encapsulated in the phrase '*from customer need to customer satisfaction through manufactured products*'. (Of course, the customer does not always recognise their own needs, so called **latent needs**, and innovative manufacture often exploits these.) This phrase embraces all of the functions of manufacturing including marketing, research, design, production, quality assurance, and financial control. This concept of manufacturing is developed in the companion *Advanced Manufacturing Design & Technology* student book and provides the rationale for large parts of the book.

Manufacturing is often divided up into sectors by the nature of the activity and the products made. The diagram opposite shows the division of manufacturing sectors used in devising a vocational course in Manufacturing. There are many generic principles that apply across all sectors of manufacturing and these are developed in the student book. This allows the book to be used by students working in a variety of materials and focus areas. The examples of products and case studies in the book have been deliberately chosen to illustrate this diversity of manufactured products and to provide points of access, interest and relevance for all students.

The approach taken in the student book is to look at manufacturing as a system with inputs and outputs, to look at company level considerations – the business of manufacturing (Chapter 2), and then to look in more detail at designing and manufacturing at the individual product and personal level. (See 'Using the student book', Chapter 7).

Manufacturing sectors identified by the Confederation of British Industry

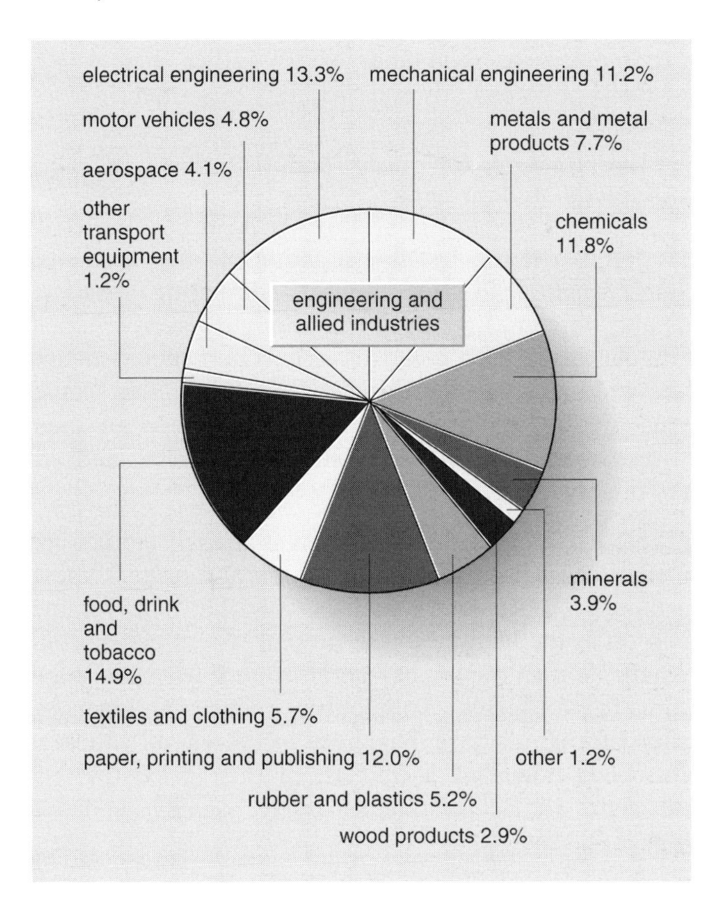

The manufacturing sectors and range of products described by an Advanced Manufacturing course for schools and colleges

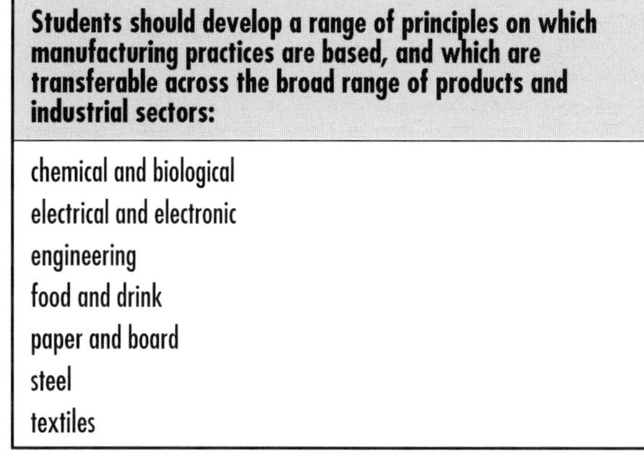

Students should develop a range of principles on which manufacturing practices are based, and which are transferable across the broad range of products and industrial sectors:

chemical and biological
electrical and electronic
engineering
food and drink
paper and board
steel
textiles

What type of activities should students work on within the industrial 'manufacturing' aspects of an Advanced course?

First, students and teachers might come to appreciate more clearly that the fruits of the students' designing and making will always be just prototypes – more or less well developed expressions of their ideas, but not fully worked-through production examples. As a result of this, these artefacts may be allowed some faults. They may be seen as a 'sketch in three dimensions', a development in dialogue with other models such as rendered sketches and technical drawings, which portrays the state of progress which the student and her/his idea had reached when, for reasons of educational management, work on it stopped. In some cases, students may take their work on to a 'fully worked-through' production prototype. They should certainly be asking themselves questions such as:

- how would I design this differently if I were planning to make 100, 1000 or 10 000?
- what materials could I use?
- would I use the same manufacturing processes?

Another activity might be designing an item for small batch production (say 20 to 200), taking into account the tessellation of parts to reduce materials wastage, the minimisation of the number of parts and assembly

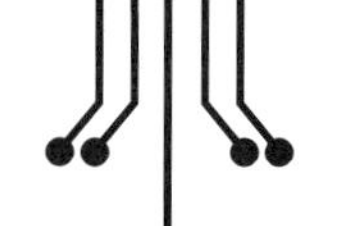

operations, the design and manufacture of jigs and fixtures, the organisation and control of production and team working demands. In other words major enterprise not mini-enterprise. This might be more achievable using such material as food and textiles rather than wood, metal and plastics for the speed, economy and flexibility they offer.

Designing for manufacture, Advanced Manufacturing, Design & Technology student book 128

Linking batch production experience to the evaluation of other students' products on the course might lead to a habit of comparing in-school constraints, methods and outcomes to those which might have pertained in a similar industrial context. 'If we could have injection moulded this product, we might have designed it like this . . . ' Certainly, this would bring to our students greater understanding of some aspects of their 'material culture' – the made world which surrounds them, and its consequences on their quality of life.

Students have been heard to ask 'Is that real or is it made?' Adults are often of the opinion that 'Mass-produced means cheap and nasty.' These two attitudes are in direct conflict, and reveal the way that high volume production has moved-on in the lifetime of the present parent generation. Mass production demands a high quality of manufacture if parts are to be interchangeable, and in a highly competitive consumer market, unprecedented quality of manufacture has been achieved at similarly unprecedented prices. Cars now cost a fraction of their price (in real terms) of forty years ago, are of inestimably higher quality, and are produced in vastly higher volumes.

It is essential that students on Advanced courses understand the means by which their society supports its needs: how things are made, why they are like they are, why they cost what they do, how they are brought into being.

For further understanding of the Project's approach to manufacturing you should look in the *D&T Routes Teacher's Resource for KS4*, pages 100–121. The InSET activity on pages 109–111 is particularly useful in understanding the implications of different volumes of production.

What do students do on Advanced courses in manufacturing, design & technology?

Assessment in these courses normally focuses both on the outcomes of the student's work and the processes used to achieve them.

The outcomes could include:

- a design portfolio
- designed and manufactured products
- evaluations of existing products and applications
- industrial case studies
- reports on various aspects of their work, such as production plans they have drawn up and used to manufacture products, reports on visits to industrial and other premises, summaries of research and investigations they have carried out
- module tests and terminal examinations.

The processes that could be assessed are:

- the quality of the planning, information seeking and evaluation carried out by the students

- how students developed key skills in areas such as ICT, numeracy and communication as part of their work
- students' logs and records of actions and decisions taken
- students' evaluation of their work and the actions taken as a result
- the quality of research carried out and how the outcomes of this research are used
- working with others both in and out of school/college.

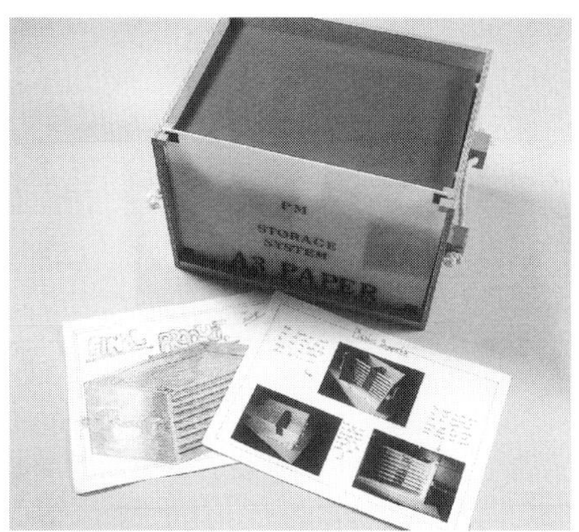

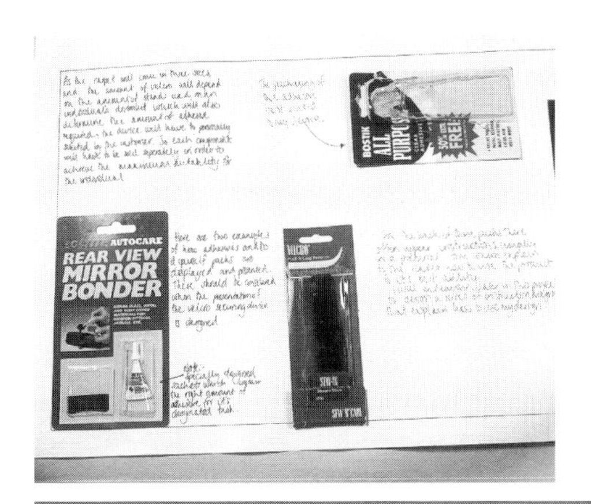

Characteristics of all students' work at Advanced level

All work at this level should exhibit characteristics at a level which significantly extends that achieved at KS4. There are six aspects by way of illustration:

1 Students' increasing autonomy and decision making

- taking more responsibility for the planning, organising, managing and evaluation of their work

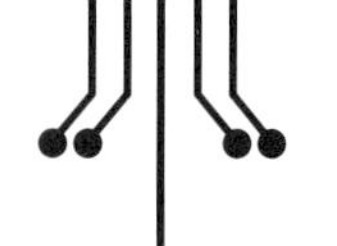

- developing and using their own individual action plans
- preparing themselves for assessment
- developing and using key skills
- making use of a wider range of people to support their work.

Chapter 1 in the *Advanced Manufacturing, Design & Technology* student book is designed to help students acquire the skills they need to take these responsibilities, building on the work established in the same chapter of the *D&T Routes Core Book*. Some post-16 students may find the latter more accessible, and this will not cause difficulties as the two books are entirely compatible in approach, offering differentiated resources.

2 A higher level of designing and manufacturing

Chapter 2 of the student book provides many insights into the quality of designing and manufacturing that is possible. One of the key features of high quality work on Advanced courses is that students always seem to be working beyond their expected level of skill and ability – they are constantly being challenged and set goals that are demanding yet achievable.

Students will acquire a wide range of designing and making skills but will develop depth in some of these related to their area of interest and the products they develop. That this depth of intensive involvement often takes them beyond their teacher's knowledge demonstrates the importance of working with a range of other people. It also requires the student to develop the skills of working more autonomously and for the teacher to adopt more of a mentoring role rather than being the source of all information. This issue is explored further in Chapter 3 **Teaching and Learning Issues**.

Students' work should reflect the complete process of manufacturing – from customer need to customer satisfaction through manufactured products. This is explained fully in the Advanced student book.

3 Working for and with clients

At this level students should be identifying clients for at least some of their projects, which might be commercially commissioned. This will allow them to develop a wider range of skills and learn about manufacturing, as they will be operating in all of the stages 'from customer need to customer satisfaction through manufactured products'.

4 Taking account of a wider range of issues

Students should be taking account of a wider range of people – user groups rather than individuals. They should take into consideration a wide range of 'values issues' – environmental, social, economic.

5 Making use of a much wider range of support

At this level students should make use of a much wider range of people to support them in their work; this will include people both in and out of school. They should be making contact with local organisations and companies who can provide help and also act as clients for their work. This will need to be managed by the teacher in partnership with the student.

6 Student-teacher-mentor partnerships

The teacher-student partnership is vital for success on Advanced courses. As has already been established, students will need to take much more responsibility for their work, while the role of the teacher changes as a result of this. Other people will be involved in supporting the students –

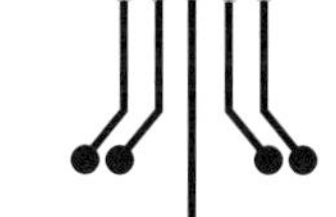

sometimes arranged by themselves, at other times organised and directed by the teacher. If this is to work then it has to be seen as a partnership. As with any partnership, this puts responsibility onto all sides but also suggests that all should benefit. Page 48 of the *Advanced Manufacturing Design & Technology* student book sets out different partners' benefits.

InSET – Setting the standard for students' work

Read the following student case studies in the *RCA Advanced Manufacturing Design & Technology* student book:
Asda Supermarkets, page 220
PDSA, page 224.

1 Note down how these demonstrate the characteristics described above.
2 Do the same for work produced by your own students.
3 Note areas you need to develop, areas where the students are not being given full opportunities to demonstrate these characteristics.
4 Over time, build up a collection of students' work that clearly demonstrates these characteristics. This can be used for developing further tasks and assignments and by students to give them an indication of the standard of work that is both expected and achievable.

CHAPTER 3 TEACHING AND LEARNING ISSUES

Styles of learning

There is now widespread recognition that learning is a complex process and all of us learn in different ways. In planning any learning activity there are a number of variables that need to be considered:

- the variety of different learning styles amongst the group of students
- the range of teaching strategies available
- your own preferred learning style and how this affects the teaching strategies you adopt
- how the approach taken is affected by the nature of the activity, the expected outcomes and where the activity will, or could, take place.

On Advanced courses there is the additional consideration that the 'teacher' may be an industrial mentor or someone else working in partnership with the students.

Many of these teaching and learning issues are developed in the *D&T Routes Teacher's Resource for KS4* including a number of InSET activities. These are not repeated here and should be referred to. Chapter 1 of the RCA *Advanced Manufacturing Design & Technology* student book also deals with this in a practical way for students to help them take more responsibility for their own learning.

You may also find it useful to read Accelerated Learning in the Classroom by Alistair Smith (published by Network Educational Press, Tel. 01785 225515, Fax. 01785 228566, ISBN 1-85539-034-5). Although this book is not focused on D&T, it explains much of the theory and suggests practical activities that could be used with students. See also: www.campaign-for-learning.org.uk/whatkind.htm

Structuring activities

One tried and tested method for delivering courses which require students to work on long assignments and take more responsibility for their work uses the following components:

Structured inputs – to provide the context for the students' work and some background information, to help with key points to be included in planning, to establish key ideas and concepts.

Focused, structured assignments to:

- equip students with skills in planning, identifying sources of information, organising their work and evaluation
- extend students skills, knowledge and understanding, and provide coverage of some important concepts in key areas needed in the assignment
- integrate key skills that will be needed, such as particular IT skills (e.g. using a spreadsheet) or mathematical skills.

Full assignments – where students take full responsibility for the complete process and to meet the requirements of their course (although these can be broken down into a series of shorter tasks for some students who need this additional level of support).

Learning activities on Advanced Manufacturing, Design and Technology Courses

Here is a description of a range of learning activities that can be used on Advanced courses in Manufacturing, Design and Technology:

- **Focused tasks** – practical, skills based, focus on a key piece of knowledge or understanding, highly structured or more open leading to a range of possible outcomes, available as support to move–on students experiencing a 'block', providing additional breadth or depth for some students.
- **Research tasks** – information gathering, investigative, including using information and communications technologies (ICT).
- **Product evaluations** – could have a specific focus, e.g. looking at values issues, design semantics, determining key scientific principles used in the design or manufacturing methods. These tasks are particularly useful as regular student-led discussion sessions.
- **Design tasks** – some not necessarily leading to made products, some leading to mock-ups (e.g. in card), some as 'design outcomes' only.
- **Design and make assignments** – prototypes, one-off products, higher volume production in groups.
- **Manufacturing assignments** – a design is given and teams plan and execute batch production.
- **Lectures and demonstrations** – by the students to their peers or to other audiences, using as wide a range of people as possible.
- **Visits** – planned by the teacher, planned by the students, general to give wide experience or focused.
- **Students working in another location** – short tasks or complete assignments, in industry, college or University.
- **Tutorials and seminars** – for individuals and groups of students, by the teacher and/or industrial mentor.
- **Simulation activity** – IT based e.g. *Virtual Factory* (from Denfords) or *Virtual Company* (Free Enterprise from ORT), or simulating a production process to understand issues of high volume production.

InSET – Using a range of learning activities

Any learning activity needs to be matched to the learning outcomes you are trying to achieve as well as to the learning styles of the students.

1. Work as a group to list some of the key outcomes required from the course your students are following.
2. Identify the learning activities that could be used to achieve each or some of those outcomes.
3. Think about the type of student who would obtain the most from each type of activity.
4. Use this to help plan a variety of activities to achieve the outcomes you require.

CPD Activity: The reflective teacher

How do you improve with practice? Developing self-awareness through the evaluation of your own work is important to improving the quality of your teaching and of building a range of skills on which you will draw in the future. The following questions may be helpful though you may wish to add some of your own.

For any particular programme of work that you are involved with:

- What were the learning needs to be met?
- How did I set out to meet these needs?
- What did I hope to achieve?
- How successful was I?
- What measures have I used to prove success?
- Were there opportunities in this lesson which I overlooked?
- Were there pressures/constraints which I could have predicted or managed better?

In the lists below cross out the terms which you feel do not apply to you. Add others which better describe your style.

Mark with an * those areas you would want to discuss with colleagues in the department, with a professional tutor or with someone from the management team in the school.

Identify the area you would choose to work on first in improving your teaching.

Attitude to students
sensitive to students' needs; flexible; appropriate; inappropriate; too rigid; overly aggressive; too friendly; too distant; disinterested; unapproachable; unhelpful; interested; responsive

Communication with students
understanding; sensitive; clearly stated instructions; unambiguous; positive reinforcement; high expectations

Relationship with colleagues
positive; helpful; willing to learn; anxious to contribute to team work; overly critical; too reserved/aggressive; unwilling to listen/learn; unco-operative

Implementation of curriculum
successfully adapts curriculum guidelines; interested in staff/curriculum development; contributes to curriculum discussions; fails to implement policies; has no clear teaching objectives; innovative

Self-awareness
reflects on teaching/learning programmes; uses advice offered positively; aware of development needs; self-critical

Acceptance of advice
open to learning; asks questions; uses advice appropriately; adapts needs; unresponsive; closed; certain of own approach; unaware of needs; obstructive.

Teaching style
Flexible → Inflexible
Controlled → Gives freedom
Planned → Unplanned

Learning activities used with students
stimulating; imaginative; productive; creative; book-based; repetitive; class-orientated; discussion-focused; interactive; cross-curricular

Questioning techniques used with students
open; stimulating; encouraging; closed; responsive; inhibiting; undemanding; thought-provoking

Classroom atmosphere
conducive to thinking and learning; clear contexts for learning; encourages interaction; develops confidence, participation; discourages interaction, discussion; inhibiting; purposeful

Classroom discipline
appropriately flexible; ordered; calm; busy, too lax; too severe; inconsistent

Class management
well organised; resources accessible; students have responsibilities; tidy; disorganised; untidy; resources missing or broken; clear areas.

Use of resources
highly effective; organised; variety of resources available; adapted resources; little used; badly managed; ineffective use; inaccessible; accessible

Identifying your own preferred learning style
It is highly likely that you will be learning as you work with students on Advanced courses and you should consider this to be part of your own professional development. This will apply to both teachers and industrial mentors. It may be possible that you can use this to contribute to a further qualification. It will help you if you can identify your own personal preferred learning style and plan future work accordingly.

The impact of new technologies on learning

After many years of computers being present in schools and much talk of impending major changes we have reached the stage where their impact on learning is really beginning to be significant. The diversity of progress in different schools is striking but the leading ones are now making real use of the convergence of information and communications technology (ICT).

Communications links can start to be effective with little investment if access to the Internet is established and, in the spirit of Design and Technology, students are encouraged to explore its value. The examples in the *Advanced Manufacturing, Design & Technology* student book show real value being obtained from World Wide Web searches of the simplest sort.

The British Educational Communications and Technology Agency (BECTa) have supported guides for IT and D&T in schools, having identified two leading areas for IT based learning support. These are modelling and CAD/CAM manufacturing. The Project's *Challenges* and *Routes* series of books repeatedly feature enhancements for activities and some discrete tasks in both these fields. Some world-leading CAM machinery is in schools and being used for highly purposeful manufacturing rather than the ubiquitous chess pieces originally seen. For 'modelling' much can be done with standard paint and draw applications as well as word-processing and DTP being the standard (over-used?) facility for report writing and presentation. The CBI's 'Manufacturing by Design' software brings parametric drafting capability with a small task folder and has been extended into the communications field with Internet links. DATA is managing the dissemination of PDC's Pro/DESKTOP advanced parametric software with training for teachers. The Technology Enhancement Programme made available from the late 90s some excellent CD-ROM based resources specifically for Manufacturing and other aspects of D&T which will hopefully become web-accessible.

Web technologies are used in schools' 'Intranets' linking applications and information banks, school or community-wide, to bring much improved accessibility to students within and outside both the school buildings and working day. D&T Online (www.dtonline.org) opened the school door to integrated purpose-made applications deliverable over World Wide Web systems, replicating industry's use of electronic product definition (EPD) supporting data sourcing, design modelling (drafting), automatic bills of parts and machining outputs (CAM), initially with disc-based software and then over the Web, if a little primitively.

The government-funded 'national grid for learning' will take some years to really meet its potential but many Internet service providers (ISPs) are now making available enhanced learning support, often with premium services at extra cost. Some of these include D&T specific material and applications. Also individual companies are very actively supporting D&T dedicated applications which allow students to experience first-hand industry-like collaboration at a distance through ICT.

Denfords remote manufacturing case study, Advanced Manufacturing Design & Technology student book 90

Given this level of activity nationally it can no longer be regarded as sufficient for a D&T department not to be making real use of ICT. The implications for school-based as well as lifelong learning are enormous. The following table illustrates some of the changes in teaching and learning that will result from the availability of these new technologies

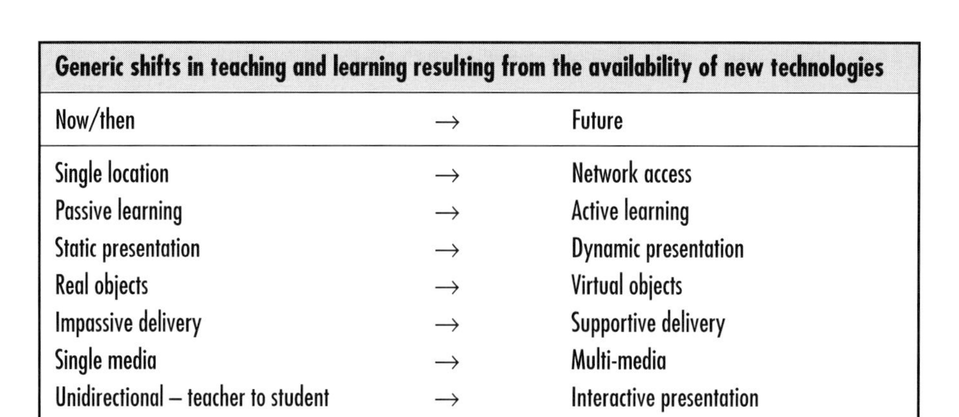

Generic shifts in teaching and learning resulting from the availability of new technologies		
Now/then	→	Future
Single location	→	Network access
Passive learning	→	Active learning
Static presentation	→	Dynamic presentation
Real objects	→	Virtual objects
Impassive delivery	→	Supportive delivery
Single media	→	Multi-media
Unidirectional – teacher to student	→	Interactive presentation
Broadcast delivery	→	Personal delivery

InSET – Planning for the impact of new technologies

Use the Now/then column in the table to review your current practice.
How far have you moved towards the Future?

What changes do you need to make to maximise the potential of the new technologies?

Increasing the student's level of autonomy

If students are to work in a more autonomous way, taking more responsibility for their work and for making decisions, there are three essential features to establish:

- **high quality and wide ranging resources** – to support the student including written, visual, IT based resources and access to people and information
- **tutoring** – provided regularly, either individually or in small groups, these tutorials could (and should) involve other people who are involved with supporting the student, for example, an industrial mentor
- **management** – to provide a clear framework for the learning to take place and to be monitored and evaluated.

Students need to develop and use **individual action plans** to help set objectives and monitor their achievements. Students will need advice and training on issues such as:

- target setting
- establishing clear expectations and outcomes for the tasks and assignments they work on and matching these to the requirements of their course
- ways of monitoring progress towards these targets and the achievement of the outcomes
- study skills
- project and time management
- how to select from and make effective use of the resources available.

Advice on these Individual Action Plans is included in Chapter 1 of the RCA *Advanced Manufacturing Design & Technology* student book.

The process established in the RCA's *D&T Routes Core Book* in the section on Teamwork should be developed. The suggestion is that students form Support Groups that meet perhaps once a fortnight. These groups can be used:

- to allow students to share their problems and lead to possible solutions

- to provide encouragement and support for each other
- as a means of obtaining feedback on ideas and plans
- to share different expertise and experiences within the group
- for evaluation purposes
- to develop team-working skills.

Some techniques for working in teams are also explained in the *D&T Routes Core Book* and in the *Teacher's Resource for KS4*, and the *D&T Challenges* Green Teacher's Resource.

Where will the students learn?

The majority of, if not all, students working on Advanced courses in Manufacturing, Design and Technology will spend some of their time out of school/college. Some students will spend more time working in companies, other colleges and elsewhere than in school. Students need to be both supported and provided with skills so that they can take responsibility for managing the *whole experience* – work in school/college, in industry and elsewhere. You should not expect students to maximise these experiences without this support and training.

Some characteristics of autonomous learning

Students have direct interactions with the resources

- This enables the student to access sources and places of learning without going through the teacher.
- It involves students in planning how to use the resources and for what purpose.
- Students develop planning, organisational and management skills and skills in the use of individual action plans.
- Students develop key skills.
- Students can take responsibility for their own learning.

The key functions of the tutorial

- To enable the students to explore and understand the objectives of the task or assignment.
- To help students set targets and develop their individual action plans.
- To enable students to negotiate learning tasks and assignments to meet the targets.
- To support the student's learning.
- To enable the students to review progress toward targets.
- To allow the students to explain their ideas and plans.
- To provide constructive criticism and advice.

A managed framework for differentiated learning

- Allows each student to work in a way that suits them.
- Enables students to be involved in the planning, organisation, management and evaluation of their own learning.
- Provides a framework to suit the needs of students, teachers and others involved with supporting the student.

InSET – Are you a teacher or a manager of learning?

When you are working with students on Advanced courses you will need to take on a number of different roles: many of these will be quite different to the traditional view of teaching. You will often find yourself in a mentoring and supporting role acting as a 'guide' to help students find their own way through activities. Your role is to manage the learning of each individual student as well as the whole group. This will also require you to manage the contributions of industrial mentors and any other people contributing to the course. If your students have learnt to work independently this will be easier.

You will find it useful to refer to the InSET activity on 'Increasing the level of student decision making' in the *D&T Routes Teacher's Resource for KS4*, pages 50–51.

Teacher roles:

- teaching the whole group – instructor
- individual tutoring – offering support and guidance
- working with a small group
- questioning – challenging
- critical friend – evaluating and monitoring
- expert – the main source of knowledge and expertise
- directing to other sources of knowledge
- leading discussions
- offering advice
- negotiating with students and planning best course of action
- assessing
- listening.

1 Which of these roles do you normally take? You might find it useful to draw up a checklist and indicate the frequency over a period of time of a variety of activities.
2 Which others could you use? When would it be appropriate for you to do so? During what type of activity? For what type of students? (See list of activities on page 19.)

Keeping records

Giving more responsibility to the students makes it even more important to keep full records and to encourage students to do so as well. Their records will help you and both sets of records will help the students understand and influence their progress.

Developing individual student action plans

The key to developing an individual action plan is the use of the following questions:

- Where am I now?
- Where do I want to be?
- How am I going to get there?
- What are my next steps?
- How successful was I?

Where am I now?	Involves a range of self-assessment techniques so that students can audit their skills and knowledge in key areas related to the task or assignment. Can be part of an individual tutoring session. Could also be built into a group tutorial.
Where do I want to be?	Students need to understand the **expected outcomes** and the standard expected for the task/assignment and turn these into **individual targets** for themselves.
How am I going to get there? What alternatives are there?	What support is available and when? How can I make best use of individual and group tutorials? How long have I got? What is the deadline? What resources are available? Where and when can I work on the assignments? Do I need any special facilities?
What are my next steps?	Planning routes through the task/assignment. Planning how to use the resources and support available. Planning a series of activities. Producing a plan and schedule (project management).
How successful was I?	Testing and evaluation strategies to use. (Throughout the student book there are activities that build up the student's Evaluation Repertoire.)

InSET – The place of tutorials in manufacturing, design and technology courses

These tutorials can take place between students and teachers and/or industrial mentors.

Tutorials can be used as:

- **a briefing tutorial** – used at the beginning of a new topic or at a significant point and should be used to clarify objectives, explain resources available, devise a strategy for the work and agree the outcomes
- **a review tutorial** – reviewing progress, students' own reports, comparing methods used by different students and effectiveness of the outcomes, interim assessment, recording achievements and helping to plan next stages
- **a discussion tutorial** – used where an activity requires the students to reflect and consider different approaches – this needs to be carefully planned
- **a coaching tutorial** – a chance to become a more traditional teacher. Needed where there are difficult points to overcome where students might struggle if left to their own devices. Best used when the students have identified the problems providing a context and rationale for the session
- **a planning tutorial** – similar to a briefing but with a wider scope than just the task or assignment in hand; could be planning a series of assignments
- **a managerial tutorial** – a business meeting to sort out arrangements e.g. for a visit.

Consider the role of each type of tutorial for students on Advanced Manufacturing, Design & Technology courses. Who else should be involved in the tutorials? This may change for different types of tutorial. Plan these into your Course Plan and ensure that the students know when they will take place.

CPD activity: Developing your own IAP

Individual Action Plans are not just for students. You can use the process described above and in the student book for your own activities. Planning in a structured way like this is a key part of continuous improvement.

Continuity and progression – the extending spiral model of learning

There is often confusion between the terms continuity and progression.

Continuity is about establishing and carrying forward key principles and the approaches taken. Ensuring continuity in D&T is a complex task, it requires logical links in experience for students as they move from Key Stage 4 into post-16 education. 'Continuity lies in the eye of the beholder'. If students cannot perceive the links between learning experiences then the links could be said not to exist. Continuity can be achieved without progression if a student has a tendency toward either repetition or regression (falling back to earlier levels of work). It is important that students develop a long term overview of their D&T learning.

The RCA Project is committed to a spiral model of learning, the central message being that areas of skills and understanding are revisited again and again but at progressively higher levels. As learning experiences continue post-16, skills and understanding are taken further and new aspects introduced such as explicitly integrating key skills and working with a wider range of clients.

Effective continuity is also vital to help students understand this model. Essentially, the same generic process and skills are employed at KS3, KS4, on Advanced courses, at graduate level and by highly experienced and skilled practitioners. The difference is the level of skill, knowledge and understanding and their application. It is vital that students retain a sense of adventure and a feeling that they are making progress throughout the key stages; their previous experiences must be valued and acknowledged. For Advanced courses, this means building on students' KS4 experiences. The diagram opposite shows the key principles that were established in the *D&T Challenges* series and carried through *D&T Routes* at KS4.

This spiral diagram starts by portraying the circular nature of a designing and making activity, beginning with an evaluation of needs and ending with an evaluation as to whether these needs have been met. Students making progress through repeated experience of this cycle can be represented as rising (in standard) from the circle to a spiral, covering the same ground, but at a higher level.

Progression: each learning activity should enable students to develop and extend their knowledge, understanding and skills in several aspects of the designing and making process. A key feature is to 'begin where the students are' – building on their previous experiences and levels of understanding/skills, and challenging them to help them to move on. In planning a sequence of activities the most important consideration is that, over a period of time, many if not all aspects of designing and making – skills, knowledge and understanding – should rise in standard.

PROFESSIONAL
Professional designers, technologists, engineers etc. do all these things on a more sophisticated level.

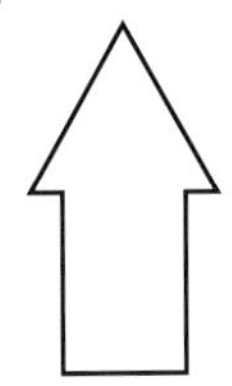

ADVANCED
Students on Advanced courses will take more responsibility for their work, work to a higher level of design and manufacture with a wider range of people, including working for and with clients, taking account of a wider range of issues.

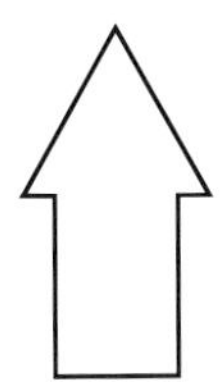

KS4
During KS4 students will often specialise in designing and making with a limited number of materials, sometimes just one. They will begin to focus on designing and making products for manufacturing and relating their work to industrial practices. They will consider applications of systems and control in their designing and making. They will self-direct their work for lengthy periods.

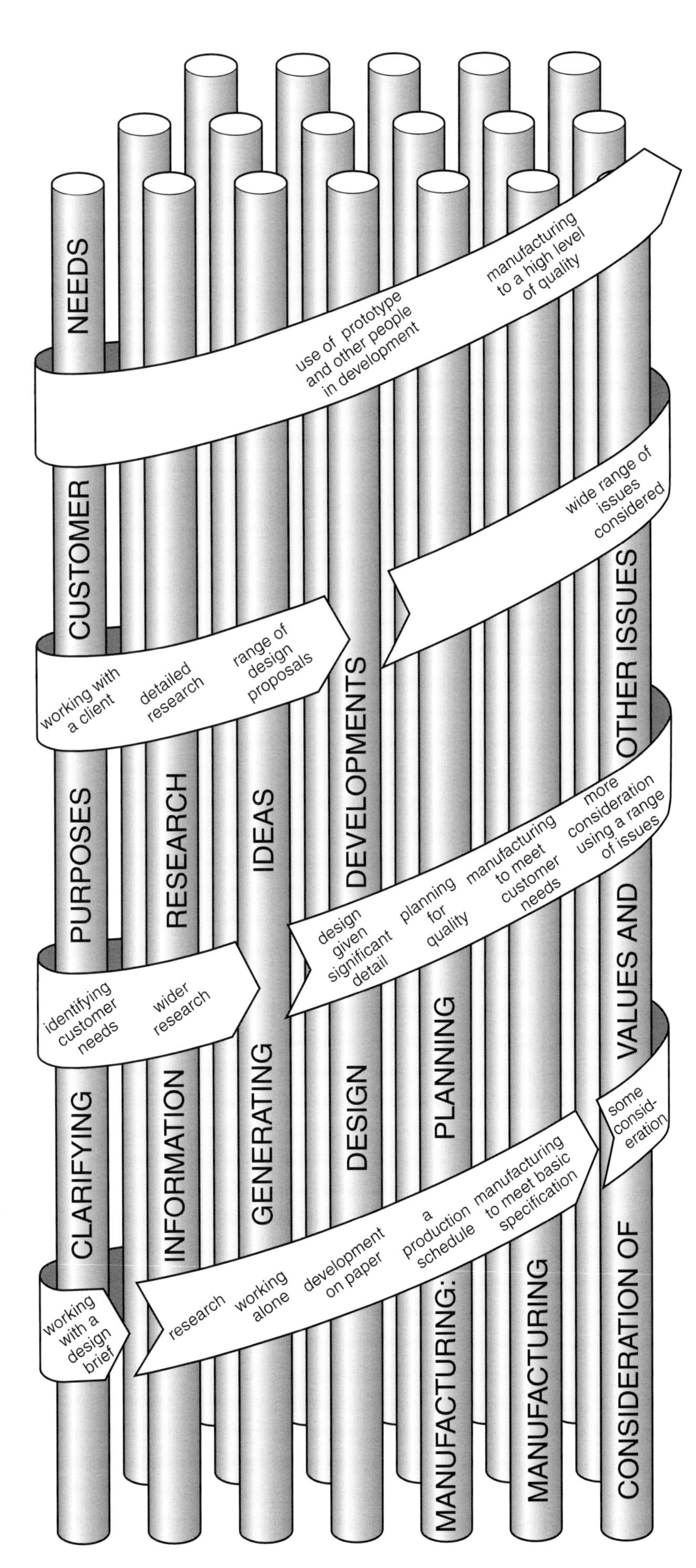

InSET – Planning for progression

When planning an individual activity it is important to be clear about its key features. For example, will it develop particular designing skills or particular pieces of knowledge?

When planning a sequence of activities, some should further develop skills and understanding developed in earlier activities. Some should introduce new skills and understanding.

Use this progression checklist to help with this planning.

Progression checklist
Progression in learning can only be defined in terms of the individual:

- does the activity allow each student to make appropriate progress?
- can each student develop increased autonomy?
- are the objectives and expected outcomes clear so that each student can set targets and monitor his/her own progress?

Progression in process skills; does the activity allow the students to:

- prepare and follow their own design briefs?
- apply existing skills and knowledge?
- develop new skills and knowledge?
- work beyond their personal experience and in unfamiliar contexts?
- be creative and take risks?
- cope with a greater number of design parameters and variables?
- use a wider range of equipment, processes, materials skills?
- evaluate against more complex specifications and use more stringent quality procedures?
- apply more sophisticated value judgements and optimise decisions when confronted with conflicting demands?
- work with customers and clients?

Progression in propositional knowledge; does the activity allow the students to:

- learn about an increasing range of materials, manufacturing processes and technologies?
- move towards accepted/industrial/professional practice?
- design and manufacture for different forms and scales of production?
- apply knowledge and skills from other sources?
- develop an understanding of existing products, user needs and preferences?
- learn how to gather information to make judgements about the impact of their designs – social, environmental, economic?

Progression in practical skills; does the activity allow the students to:

- learn and apply a wide range of skills?
- increase their skills development – from competency through to mastery enabling them to design and manufacture high quality products?

Differentiation

Having lived different lives for at least 16 years, students on Advanced courses will have different levels of ability, different aptitudes and a different range of experiences and backgrounds. They are also likely to have different expectations of the course and different aspirations. This means that the issue of differentiation is equally, if not more important, on Advanced course than in earlier years.

Differentiation means sufficiently and appropriately challenging each and every student:

- building on their previous experiences and understanding
- developing their understanding and skills to their full potential
- providing learning experiences that are appropriate to the student, taking account of their preferred styles of learning.

The **Introduction**, **Intervention**, **Extension** and **Enrichment** (IIEE) model of differentiation established successfully by the RCA Schools Technology Project in KS3 and KS4 can be applied equally successfully on Advanced courses.

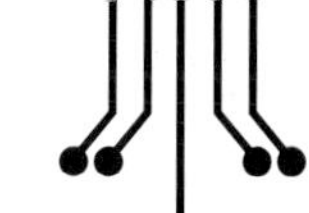

The principles of the IIEE model

When planning a teacher–led activity, such as is common in Y12 modules, it is essential that you plan ahead for differentiation. Ask yourself the minimum, the normal and the highest level of result you might realistically expect. Ask yourself how you will identify those who might need to work differently to the majority in a group, for example:

- faster at a more sophisticated level
- slower but to a sophisticated level
- with special skills but special limits/weaknesses
- faster but with less attention to some aspects.

Then consider how the work might be varied to accommodate each of these. Note that all students, for reasons of class management, may begin and end at the same time despite having had significant differences in their experience.

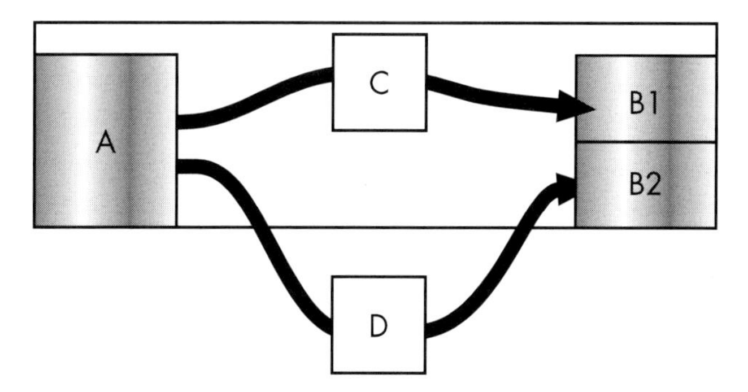

The IIEE model in action. A represents the starting point for the activity, B1 the end point for one student, and B2 for another student (working faster and completing a further extension, D). C represents enrichment – a subsidiary activity completed to a better level.

Introduction

The introduction to a class activity, planned and delivered by the teacher, is crucial. At this stage interest and motivation may be established, or may not. It is not axiomatic that certain activities appeal to and interest all students. With an introduction that is convincing – that convinces the students that it is worthwhile – a project may have a flying start. This places a responsibility on the teacher to resource it well and deliver it with enthusiasm. It also requires early consideration of ways in which the appeals of the project may be directed to students with different interests. Do not assume that all of them are interested in the same things as you!

Intervention

An important aim of designing and making assignments at this stage is for them to be more student-directed. Without 'ownership' it is merely an exercise. It follows therefore that the teacher's aim is to be 'hands-off', to allow the student to take the lead in determining her/his direction and working method, even when the assignment has been set by the teacher. The teacher's role then becomes one of choosing when to intervene to suggest re-direction for the student, having identified a need to steer her/him more profitably. As this will involve a conversation with the individual, this is a fundamental opportunity for differentiation throughout the intervention. This differentiation can be achieved through: style and level of discourse; complexity of demand assumed; sophistication of thinking engaged in and level of difficulty inherent in the direction the student is advised to go in.

The intervention can take place informally or in a more structured way through tutorials.

Extension and enrichment

The teacher's aim with able students is to extend the work for some and to enhance the work for others. Extension tasks expect some students to do more parts of a project, carrying out activities others do not. Enrichment refers to some students doing standard activities that others are involved in more thoroughly or at a more sophisticated level. Neither of these need take more time but require careful management by the teacher and a good knowledge of the students. To simply think in terms of norm plus extension (as many teachers do) is not adequate in D&T at this level.

Either extension or enrichment requires it to be accepted as normal by students that different members of the same class work to different targets. They are familiar enough with this in the case of differentiation by outcome – they will have seen the results of fellow students who go far beyond the achievements of others. However, it is different if this is formalised with expectations agreed at the start of a project that some will go well beyond the norm (and some stay well within it).

A tutorial approach (page 25) is particularly effective in managing differentiation. Group tutorials can be held with small groups with common needs.

InSET – Planning for differentiation

Planning for differentiation involves planning for:

- different styles of learning
- different prior experiences of students
- different abilities and aptitudes
- the different needs of different students
- making the best use of the resources available.

For each activity you need to expect a range of differentiated outcomes.
These need to be communicated to students so that they can set their own learning targets.
Record with colleagues how each of these can be planned for in one Y12 module.

Assessment issues

Assessment on Advanced courses requires consideration of three key issues:

1 Preparing students for summative assessments such as examinations.

2 Formative assessment to monitor and review students' progress, using this information to help them set targets and goals.

3 The role of students in assessing their own and each others' work and monitoring and reviewing their own progress.

Refer to the assessment sections of *D&T Routes Teacher's Resource for KS4* for more coverage of issues related to assessment. In addition, Chapter 1 of the *Advanced Manufacturing Design & Technology* student book (page 16) covers some issues related to students' self-assessment which are developed here.

The basic principles of student-centred assessment

Assessment is a vital part of the learning process; this includes self-assessment by the students and feeding back the outcomes of assessments made by the teacher or others.

The underlying principles that underpin student-centred assessment are that:

- students will develop and progress and their abilities are not fixed
- the aims and objectives of an activity and the planned learning outcomes are shared with the students
- deadlines are negotiable whenever possible, and agreed by consensus
- the assessment should allow students to build on their achievements and overcome weaknesses when it is important to do so
- the outcomes of assessment are used for setting targets
- students receive regular feedback, including feedback from their peers
- students need to learn how to reflect on and review their work and their learning.

InSET – Planning for assessment

When planning any student activity consider the three key issues detailed opposite.

Identify:

- those aspects of the activity and outcomes that are appropriate for the students to assess for themselves: their own learning and their development
- the formative assessment techniques that you will use to enable you to monitor progress and feedback to students
- opportunities for these assessments to prepare students for summative assessments.

Build this into your planning process.

Values issues

Including a consideration of wider values issues is a feature of all of the publications of the RCA Schools Technology Project. On Advanced courses students should be encouraged to develop a critical thinking approach to their work incorporating a range of values issues including:

- the benefits and disadvantages to the users of products and others
- consideration of the impact of products and technologies on a much wider range of people
- the principles of inclusive designing
- a respect for products and design issues from outside their own culture
- evaluation of environmental impact – including life-cycle assessments, intended and unintended outcomes
- the use of social auditing alongside financial auditing
- the use of technology assessment
- consideration of ethical and moral issues
- economic issues.

These issues are covered in the *Advanced Manufacturing Design & Technology* student book and you will also find useful information in the *D&T Routes Core Book*.

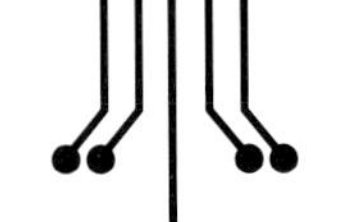

InSET – Developing progression in values issues

Use the information on page 37 of the *D&T Routes Core Book* to establish what students may have experienced during KS4.

Check this against the actual experiences of your students.

Identify those with especially useful experiences to contribute.

Develop a list of ideas for incorporating values issues into assignments.

Check the assignments you intend to use to find opportunities for including these ideas.

Making value judgements	
Your response to the product	What is your initial reaction to the product?
	Would you like to have it? Why?
	Who do you think it is intended for? Why do you think this is so?
The 'need' for the product	Why is the product needed?
	How do you think that the needs were identified?
	Who will benefit from the manufacture of the product?
	How will they benefit?
Design	Who made the decisions about the design?
	Is there a choice of designs?
	How do you think the design was developed?
	What materials are used?
Manufacture	How are these materials obtained?
	What other resources are used to manufacture the product?
	What impact could using these materials and resources have on the environment or on other people?
	What happens to any waste produced during the manufacturing process?
	What are the working conditions in the manufacturing plant like?
Promotion	How is the product promoted and packaged?
	Who is the promotion aimed at?
	What assumptions have been made about these people?
	Does the product have an identity or image?
	How has this been achieved?
	Does this cause any offence to some people? Why?
Use	Will the product have any impact on the environment?
	What effect could it have on other people?
Disposal	Does the product contain any reusable, renewable or recyclable parts?
	What will happen to the packaging?
	How will the product or any waste left over be disposed of?
Values:	List these values in the order of importance to potential customers for the product.
technical	How would the order change for people of different age, culture or
economic	lifestyle?
environmental	Put the values in the order of importance to the manufacturer.
social	Put the values in the order of importance to the customer.
aestheric	List the values in the likely order that would be used by an
moral	environmental and a social pressure group.

The place of key skills in Manufacturing, Design and Technology courses

There is now widespread agreement about the importance of Key Skills, what these are and their importance in making people employable in the short and long term. There is an increasing recognition of the need for everyone to embrace the idea of lifelong learning – this is recognised by initiatives such as the National Record of Achievement (NRA) or Progress File, the Royal Society of Art's Campaign for Learning and the setting up of a University for Industry. These Key Skills are also an essential part of any Advanced course and are generally taken to include:

- communication
- application of number
- information technology
- improving own learning and performance
- problem solving
- working with others

Metacognition

Probably the most important skill in lifelong learning is to learn how to learn; this we do through developing metacognition. We should constantly expect students to consider why they are doing what they are doing, and periodically to look back and reflect on how they did it and whether they might have done it any better.

Developing key skills

Students should be:

- provided with opportunities to develop these key skills in the context of their work in manufacturing, design and technology
- encouraged to include them in their own action plan (page 25)

Many of these skills are fundamental to successfully working in an autonomous way and will be better developed by students given this responsibility.

Chapter 1 in the Student book provides guidance to students on how to identify opportunities to both develop and make use of these skills (page 18).

InSET – Auditing and planning key skills development

In developing tasks and assignments you need to identify explicit opportunities to integrate key skills. You also need to ensure that these are developed progressively. Students will also be developing these skills in other activities they are involved in; this also needs to be taken into account – much of the responsibility for this should be taken by the students.

In the three core skills areas (communication, application of number and IT), identify the information available to you from your students who are concluding KS4. If formal assessments of key skills have been made in, for example, Part One GNVQs, then this will be an important start, but also discuss with colleagues:

- how should you obtain this information?
- when should you examine and make use of it?
- what use should you make of it?
- how can it help you monitor and develop their skills?
- how to compare input and output summaries to feed back into your course planning.

CPD Activity: Auditing your own key skills

Continuing to develop key skills is an essential part of your lifelong learning. You may find it useful to develop your own individual action plan. The National Record of Achievement or Progress File is intended for all – it is designed to be a tool to support lifelong learning – so you can use it for yourself. Information on the Campaign for Learning, which supports lifelong learning, is obtainable from The Royal Society of Arts, 8 John Adam Street, London WC2N 6EZ (www.campaign-for-learning.org.uk).

Consider how your skills in these aspects of your life could be further developed:

- communication
  - – with students
  - – with colleagues
  - – with your managers
- applying number analysis skills in monitoring the progress of
  - – your students
  - – your department
  - – your effectiveness
- and the other four key skills listed on page 33.

Advanced studies as part of lifelong learning

Students' work on Advanced courses should be seen as part of their lifelong learning. One way to make this successful is to encourage the use of the National Record of Achievement or Progress File, which is intended to be a lifelong process. Information about these will be available in your school, college or from QCA. The ASDAN Award scheme is also useful.

The ASDAN Award Scheme

The broad aim of this scheme is to offer an activity based curriculum for those of all abilities within the 14–25 age range, together with a framework for assessment. It is based on the development, demonstration and accreditation of personal and social skills within a variety of contexts. The key areas are:

- improving own learning and performance
- communication
- working with others
- problem-solving
- application of number
- information technology.

It is designed to support both the formative and summative processes of the NRA or Progress File.

The accreditation system operates from Entry level up to University entrance. The Universities Award is included in the UCAS guidance to applicants under 'Qualifications' and includes key skills.

For further details contact:
ASDAN Central Office, 27 Redland Hill, Bristol BS6 6UX or see www.asdan.co.uk.

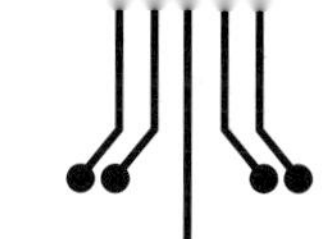

CHAPTER 4 PLANNING A COURSE

The *D&T Routes Teacher's Resource for KS4* establishes some basic principles for turning a syllabus or specification into a complete course. These principles also apply to Advanced courses and should include everyone involved with the course such as industrial mentors in a local college supporting part of the course.

Planning a complete course

Clearly, a syllabus or specification is not a course. A GCE A/AS specification lists the content to be covered and the assessment procedures to be used but it does not state how the content should be taught. GNVQ specifications set out required learning outcomes through expected Assessment Evidence. Whole course planning should take into account the planning for continuity, progression, differentiation, assessment and the development of key skills described on previous pages.

Planning should take place at three levels.

Planning level	Level of detail
Long term: planning across the complete course and year by year	Key features of long term planning: • develop a scheme of work to ensure coverage of the syllabus/specification requirements • develop continuity from KS4 and other previous experiences • develop a range of 'progression maps' for key aspects of the course to provide the framework for developing projects • build in the development of progression through key skills • ensure that sufficient opportunities for collecting evidence about students' achievements are provided • provide a balance of teaching and learning styles • develop a range of different experiences for students • plan links with other areas of the curriculum • make effective use of links with industry and other external agencies • make effective use of industrial mentors. This will also help with the management of staff, resources, equipment and rooms.
Medium term: planning assignments	Produce a plan for Y 12 unit assignments which sets out: • the learning objectives • the expected learning outcomes • strategies that can be used to ensure that the learning outcomes are achieved. This should be a source of teaching and learning strategies, not too tightly defined stifling innovation and individual approaches • strategies for differentiation • resource requirements • links and references to other projects • the key points where tutorials will take place.
Short term: lessons/small groups of lessons/weekly plans 'Turning planning into good lessons'	Detailed lesson planning and associated records to ensure effective teaching and assessment: • to take account of what you find out about students' progress • to plan how students will produce the evidence they need to present • to plan the next steps in each student's development.

InSET – Working with a syllabus or unit specification

A new syllabus or specification can seem complex and difficult to follow at first. One way to rationalise the process is to break it down into manageable parts. Simplifying it and putting it into a visual form helps to identify the basic components. Once an understanding of the basic structure of the syllabus has been reached, you can start to build a coherent course.

STAGE IN THE PROCESS	NOTES
Analyse	Coursework and assessment components Course units Turn into a visual form, e.g. a chart
Rationalise each	Once the basic structure has been understood you can overlay specific topics, knowledge, understanding and skills in a progressive manner.
Structure:	
• teaching time	The organisation of staff and their time commitments can be beneficial to the flexibility of a course. Teamwork broadens your skills and knowledge base and provides opportunities for combinations of knowledge, skills and materials in projects.
• learning time	Advanced students will learn in many different environments – in school/college, on industrial placements, visits, in their own time etc. Careful account needs to be taken of this during planning. You also need to consider the time for exams/module tests which reduce contact time and exploit opportunities such as vacations which may allow time for extended research and other activities.
Resources	Plan the management and organisation of resources: • rooms • equipment • teaching staff • industrial mentors • students working in other places and with other people • consumable materials • maintenance needs • ancillary staff • learning materials • IT facilities.
Networking	There is a big network willing to help if you make the effort – students, parents, community, industry, further and higher education. Networks can promote student learning, links with manufacturing and other industries, student skill development and staff development.

Managing the sequence of work

When planning the sequence of work for students through an Advanced course, there are a number of issues to consider. These include:

- whether to closely follow a unit's structure and whether to integrate two or more units
- whether to run two units in parallel (e.g. coursework and a case study)
- using progression maps to show the main stages in progression through the course, especially the relationship between units
- using a series of shorter designing and making assignments with supporting focused practical tasks to develop key skills, knowledge and understanding – building on the RCA KS4 model
- building in opportunities to develop skills in project management during Y12 for longer coursework assignments

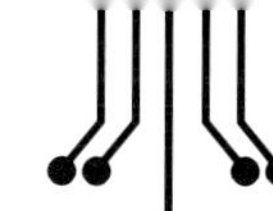

- allowing sufficient time for students to produce work of high quality
- flexible deployment of staff and resources
- opportunities for teamwork for both staff and students
- opportunities for cross-curricular links particularly with Art, Business, Mathematics and Science
- developing links with further and higher education and industrial companies, providing sufficient opportunities for students to take control of their own learning.

An example of how to produce a plan for a unit

Projects/Assignments/Activity	Key features	Time	Notes
1. Industry project. Using mass produced sound effect PCB to make batch of novelty items	To imitate key aspects of project and time management and production schedule	2 weeks	Take students through examples of how to produce a project plan, production schedule. (Use chapter 1 student book)
2. Key skills analysis	Self evaluation and producing an individual action plan	2 hours of IAP.	Show student's example (Use pages 18–20 of student book)
3. Team building session with industrial mentors		$\frac{1}{2}$ day	To be done in conference room on industrial premises
4. Introduce first student project			

Planning projects and assignments

The main features of good projects and assignments are:

- a clear assessment framework given to students so they can plan their work to meet its requirements
- the designing, making, manufacturing and other skills that can be developed by the assignment will be identified
- these skills will be built into clear strands of progression
- there will be opportunities for students to develop different outcomes to allow for their own capabilities, learning targets and creativity
- the assignment will be supported by differentiated focused tasks – some practical, others researching, investigating and evaluating products, knowledge based etc.
- links are made to the work of professional designers and industrial manufacturing practices, opportunities to consider values issues are ensured.

Using an Assignment planning grid

The grid shows one way of planning unit assignments. It enables:

- the key features of the assignment to be highlighted to help with matching the assignment to progression
- milestones in the project/assignment – key points where the student and/or the teacher needs to check progress; these will also be points that students need to build into their project management plan
- you to identify and plan the focused tasks that will be required by some or all students to develop particular skills or understanding – these will come from the Advanced student book or from other sources; these will be tasks that will move-on the assignment at that point
- links with case studies in the student book to be planned – references to case studies that will inform the assignment at that point.

InSET – Developing your own assignments

Evaluate the Assignment Planning Grid opposite in a group as follows:
Take two existing assignments and use the grid to identify their strengths and weaknesses.
Use the completed grids to improve progression through a series of assignments
Decide on the changes in emphasis needed to improve students' progression

An example of a progression outline for understanding quality systems in manufacturing

Progression – understanding quality systems	When
Quality control	
– understanding quality standards, tolerances and inspection	module 1 wk 3
– use of sampling	module 2 wk 5
Continuous Improvement	module 1 wk 4
Customers and quality	
satisfying and exceeding customer needs and wants	
– right first time, every time	module 1 wk 4
– quality assurance: ensuring quality and reducing the need for inspection	
– more complex sampling techniques	module 4
– statistical process control (SPC), control charts	module 5 wk 2
Total Quality Management	module 2 wk 6

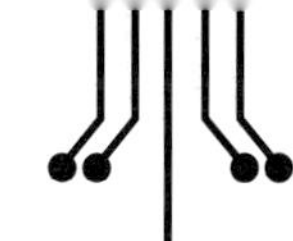

Assignment Planning Grid

KEY FEATURES: ASPECTS OF DESIGNING AND MANUFACTURING THAT THE PROJECT DEVELOPS	KEY STAGES IN THE ASSIGNMENT/PROJECT (MILESTONES)	FOCUSED TASKS	CASE STUDIES	MAKING USE OF OTHERS, VISITS ETC.
Assignment 1				
Assignment 2				

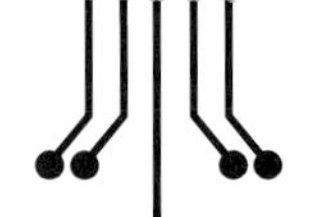

An example of an Assignment Planning Grid

KEY FEATURES: ASPECTS OF DESIGNING AND MANUFACTURING THAT THE PROJECT DEVELOPS	MAIN STAGES IN THE ASSIGNMENT/PROJECT (MILESTONES)	FOCUSED TASKS	CASE STUDIES	MAKING USE OF OTHERS, VISITS ETC.
Assignment: An Electronics workstation (This was a teacher initiated project designed to cover specific aspects of the syllabus and to illustrate the complete process from customer need to customer satisfaction)				
Discussing a brief with a client	Clarification of the brief leading to clear statement			Local company acting as client, use of visits to factory and engineers to school
Product research	Identification of the equipment required for electronic workstation	Observed use of existing stations and interviewed operators		
Establishing an initial specification and agreeing this with the client	Initial specification			
Developing a specification through focused research, initial ideas and identifying constraints		Material selection, ergonomic study	ASDA (Ch. 5)	
Making design choices judged against specification and negotiating with client	Design specification Design ideas in a form for presentation to client	2-D models, CAD, prototype patterns		
Feasibility study, testing, research analysis	Outcomes of feasibility	Production feasibility study, costing, skills audit, setting quality standards		
Development and modifications				
Final design proposal	Final design			
Pre-production prototype	Prototype with times for operations evaluated, used to plan production sequence			
Planning manufacture, costing, quality assurance and preparing a manufacturing specification	Manufacturing specification, production plan, costings	Team planning for production run; quality systems	PDSA (Ch. 5)	
Production, in-process modification, quality control	Product – production line run; 30 items produced			
Evaluation/testing	Evaluated with client outcomes of testing and evaluation			

Course: GNVQ Manufacturing (Advanced)

Parts of the course coverd by that project/assignment:

1 Process operations

2 Production plan including costs etc.

3 Work practices

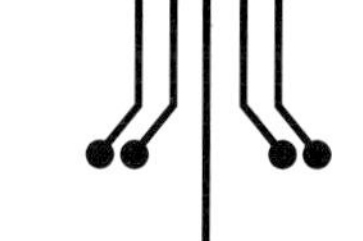

CHAPTER 5 MANAGEMENT ISSUES

Managing your resources

The range of resources available to support students include:

- learning resources
- rooms and equipment
- people and organisations – inside and outside of school/college.

D&T Routes Teacher's Resource for KS4
Resource Management 37

The most effective way of managing this range of resources is to use a Cost-Time-Resource (CTR) analysis for each major activity on the course. This process is also explained in Chapter 1 of the Advanced student book and should lead to students planning their own resource requirements more effectively.

An example of a CTR analysis for part of a course

COST-TIME-RESOURCE sheet	
PROJECT: industrial simulation – use of control systems in manufacturing	
SCOPE (explanation of activity) Students work in teams to assemble a production process. Each student takes one function and responsibility for some part of complete system	
PROJECTS/ACTIVITIES that must precede this one Industrial visit	**PROJECT/ACTIVITIES that cannot take place until this one has been completed** Industrial study
DURATION: 4 weeks	**Which part of course** Y12 Spring term
RESOURCES **1. MATERIALS AND EQUIPMENT** Polymek – partly consumable Construction kits	 Electronics and pneumatics kits Computer and interface
2. ROOM/FACILITIES Access to control room and computer **3. PEOPLE** Andy Froggatt from Technical Products Ltd – visits to school need planning **4. OTHER** Arrange trip to factory – me to plan then students May need follow-up visit to check details and get specific information	
PERSON RESPONSIBLE: A.N. Other	**DATE:** 4/11/2001

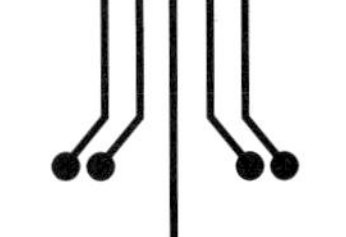

Managing your managers

Not all senior management teams have any depth of appreciation or commitment to Advanced Manufacturing or Design and Technology. However, your managers should be considered as a valuable resource to support the work you are doing but to achieve this may take some effort on your part. Establishing good relationships with your managers can lead to:

- enhanced provision and resources
- more timetable time
- higher status for the subject in the school/college leading to more students.

Poor relationships can, of course, lead to the opposite effect!

Some tips for managing your managers:

- plan regular meetings and be proactive – arrange the agenda to suit *your* needs
- produce a course handbook and make sure a copy is provided for all appropriate managers
- involve them in team meetings whenever possible and make sure the agenda contains items of relevance
- give them a copy of this book, the Advanced student book and the KS4 Routes books
- make sure they are aware of high quality work produced by the students
- provide summaries of national developments to keep the profile of the subject high
- invite them to special occasions e.g. review tutorials, displays etc.

Managing quality

There are a number of aspects of quality that need to be considered and systems should be established to monitor and manage the level of quality in all of these:

- course planning
- students' work and the quality of their learning
- teaching resources used – texts, task/assignment sheets, IT, equipment, rooms
- tasks and assignments used
- formative assessment strategies used to monitor students' progress and achievements and to set targets for future work
- support within the department and within the school/college
- support from other people and organisations.

Total quality

The concept of **total quality** within organisations embraces the philosophy of **continuous improvement** and is built on the assumption that:

- what an organisation achieves is the result of all the activities that take place within it
- it is always possible to improve what is currently being achieved
- all functions and all people involved need to be actively involved in the process of working towards improvement
- it is necessary for an organisation to collect, analyse and make use of appropriate data about the current state of affairs in order to assist the process of improvement

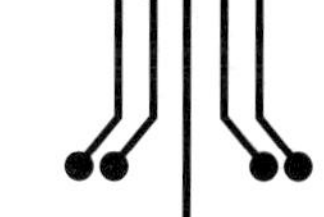

- setting targets and standards for performance at all levels can provide a useful framework to support improvement.

A drive towards total quality therefore needs to work at all these levels:

- establishing and communicating a corporate mission
  – 'What do we intend to achieve?'
  – 'What do we stand for?'
- as part of the overall strategic planning and target-setting within the organisation
  – 'How we will get there?'
  – 'How we will know when we get there?'
- at the level of each working group – clarifying operations, procedures and control systems
  – 'How do we intend to achieve what we are committed to achieve?'
  – 'How will we know that we are getting there?'

How can the ideas behind the concept of total quality work for the Advanced course teaching team?

1 **Fitness for purpose**
The idea that all people in the team need to have a clear sense of what the project as a whole is striving to achieve; reaching a precise definition of what they are trying to achieve: 'Why are we here?' 'What is our core purpose?' 'What are our core values?' 'Why do we do that?' 'Why do we do it that way?' 'Is there a better way?' The team needs continually to refine and refresh its sense of purpose.
2 **The concepts of customer, consumer and supplier**
These concepts are closely linked to the process of clarifying purpose: 'Who is this for?' 'What is it for?' 'Why are we doing this?'
An exploration of fitness for purpose needs to take into account the existence of both internal and external customers, consumers and suppliers. Who are the customers for and consumers of your teaching?
3 **The search for excellence**
Meeting – or exceeding – the customer's requirements. Improving quality requires the team to set standards and targets for excellence which allow all of the individuals and teams to assess the extent to which that desired quality of outcome has been achieved and to aspire to go beyond the original intentions to achieve further improvement.
4 **The need for systems of monitoring and evaluation**
These systems relate to the sense of core purposes as well as to the achievement of specific short and medium-term targets. Total quality entails working in a data-rich environment, knowing what we do, how we do it and what its outcomes are, at both a personal and a collective level.
5 **The need for a collaborative culture**
A culture in which people develop the skills to work in teams to achieve continuous improvement, even with limited resources.
6 **The concept of the learning organisation – the learning team**
A team which is ready to be questioning, opportunistic, flexible in the face of change; which is working towards openness and trust and seeking shared values and purposes; in which project leaders recognise that one of their major functions is to support and encourage their teams in the development of the skills implied in **total quality management**.

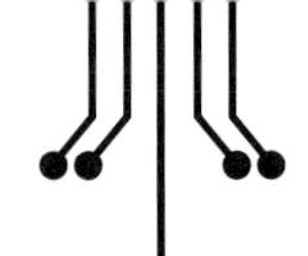

The key features of quality systems

Quality control (QC) can be summarised as *'techniques for checking quality against set standards, or within tolerances'*. It involves the inspection and detection of products not up to standard. The inspection takes place after the event – after the relevant stage in manufacturing or when the product is complete – and is carried out by trained inspectors.

Quality assurance (QA) is a series of planned and systematic actions and procedures designed to ensure that the product or service meets the quality standards. It takes place before, during and after and the aim is to prevent failure: right first time, every time. QA is the responsibility of everyone; it is a process used to build quality into every stage in the product development and manufacturing process.

Total quality management (TQM) extends QA creating a **quality culture** where the aim of the organisation is to **delight its customers**. TQM embraces a philosophy of continuous improvement where the aim is to continually improve the performance of the organisation and its products and services – always trying to make things better.

InSET – Improving quality

Use the idea of **total quality** to identify a few key quality issues in the work of your Advanced course team.

For each of the quality issues to be managed identify the most appropriate quality system to be used. How can that system be used to monitor and improve the level of quality?

Managing the requirements of different courses

It is possible that within a teaching group you may have students working on different types of manufacturing, design and technology courses, for example, working across different focus areas, working on academic and vocational units. Additionally, in many schools, because of group size, students on the first and second years of a course may have to be taught in the same group.

This task becomes easier to manage if students have been given the opportunity to work more autonomously and are used to setting their own action plans. The role of the tutorial session becomes even more important in this situation.

The key issues are:

- planning each course in full
- identifying matches or overlaps in units of work
- identifying any necessary sequences of learning, and what might come at any stage
- planning tasks, assignments and units in detail to fit the overall course plan and to meet the requirements of each course
- looking for overlaps and common areas within units – this could lead to common tasks/assignments that meet the requirements of more than one course
- negotiating each student's route through the tasks and assignments
- setting up tutorials/seminars at key points for each individual or group of students.

CHAPTER 6 DEVELOPING SUSTAINABLE PARTNERSHIPS

Most schools/colleges will have developed partnerships with local companies and other organisations and individuals; these partnerships can be invaluable in supporting students on Advanced courses through working with a wider range of people. Many of these partnerships are dependent on key individuals and collapse if these individuals move on. This section provides some advice on making these partnerships more long-lasting or sustainable.

A key to this sustainability is developing a shared vision:

- understanding each other
- understanding each other's needs
- understanding how you can help each other.

Some basic principles:

- All members of the partnership have a mutual interest in seeing young people succeed. The young people of today are the source of innovation and the drivers of future developments.
- Combining the strengths of education and industry to provide better educated young people. Everyone has something to gain from the partnership – students, teachers, schools, industry, people in industry.

Understanding each other – tackling some prejudices

Common prejudices found in industry are that people in education are:

- out of touch
- anti-competition
- not results-orientated
- intelligent and highly educated.

Also, that education as a whole is:

- undermined by short working hours and long holidays
- ineffective: does not provide young people with the skills we need
- bureaucratic: erecting barriers to effective partnerships
- under-resourced
- over-worked
- hampered by inadequately trained teachers.

Common prejudices found in education are that industry is:

- exploitative
- dirty
- based in out-of-date factories
- rife with disputes
- staffed by low-achievers
- only interested in short-term gains
- obsessed with profits
- well-trained
- well-organised
- well paid.

Clearly, these attitudes are full of inconsistencies and can only be improved by developing a much deeper understanding of each other's needs and, most importantly, the needs of the students/employees and how these are best met through partnerships. By focusing on the potential benefits to all involved it is possible to establish an effective middle ground to the mutual benefit of both schools/colleges and companies. This includes the benefits that will accrue for:

- students' learning and career progression

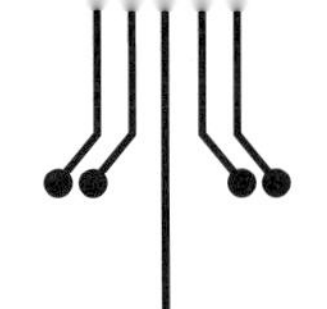

- the professional development of teachers
- the professional development of people in industry.

Understanding each other's needs

1 **The needs of industry**
Industry needs employable people; the characteristics of these people are that they are:
- equipped with a range of key skills
- adaptable and flexible
- aware of the need, and prepared for, lifelong learning.

2 **The needs of schools/colleges**
- help with meeting the objectives of courses, particularly relating to manufacturing issues, industrial practices, reflecting the world of professional designers etc.
- help with the professional development of teachers, for example, in ICT skills
- greater understanding of the needs of industry, for example, developing key skills and flexible, adaptable young people
- establishing the current and longer-term priorities.

3 **The needs of students**
These can be grouped as:
- short term – school/college course focused
- medium term – employment post-19, Further and Higher education
- long term – future employment, lifelong learning, key transferable skills.

Who can help?

There are a wide range of organisations which can help develop and sustain partnerships such as: Training and Enterprise Councils (TECs), Education Business Partnerships (EBPs), SATROs, Chamber of Commerce, your local SETPOINT and school sponsors.

The benefits of partnerships

Questions teachers ask	Some of the key benefits
Why should schools/colleges recognise industry?	• Students will spend their working lives in an industrial society. • An understanding of industry as vital to wealth creation and improving the quality of life. • Increase students' personal development and key skills. • Windows onto the world outside school or college.
What can an industrial partner offer to the school?	• Industrial mentors to work with students. • Insights into manufacturing. • Help with specific parts of the syllabus. • Visits. • Skills and resources.
Is this going to be another burden?	• Planning is essential and this takes time but will ultimately enhance the resources and teaching capabilities of the school/college. • Collaboration with industry often leads to new ideas and better ways of doing things.
What are the motives of industrialists who want to work with schools?	• Public relations. • Interest in improving the abilities and attitudes of students. • To find out more about education. • To be of service to the community.

Questions people in industry ask	Some of the key benefits
What benefits can we gain from working with schools/colleges?	• Contributions to staff training and professional development. • Provide an interest to employees which may lead to enhanced motivation and job satisfaction. • Public relations – contributions to the community. • Better trained and prepared workforce for the future. • Finding out about education now.
What aspects of industry do schools/colleges want to know about?	• The complete process of manufacturing – from customer need to customer satisfaction. • Specific aspects appropriate to the course. • The nature of the business and any issues that arise from this.
What can we offer?	• Specific skills, knowledge and resources. • Visits by students to the premises. • Visits by employees to the school/college.
What kind of people do we have who could offer something to the school/college?	• Every kind – employees at all levels can become involved and benefit in terms of personal development.
Will it take up a lot of time?	• This is something that needs to be planned carefully with the school/college and will help both in identifying the benefits and the time that can be allocated to maximise these.

InSET – Collaborative training and development

Working across education and industry – developing a shared vision.

Session 1

Working separately, answer the following.

Industrial mentors

- What is it like receiving students into your workplace?
- What are the advantages and how can these be built on?
- What are the problems and how can these be reduced or overcome?
- A student comes to you in the workplace bringing one page of information for you. What should it contain?
- When you go into a school what information do you want to be armed with?
- What information do you want to provide for the school?
- Which parts of this could be included in a joint guidance document?
- How can partnerships be made sustainable? i.e. more secure than contacts between individuals.
- How can this work contribute to your continuing professional development?

Tutors in schools/colleges

- What is it like receiving people from industry to help you and your students deliver the course?
- What are the advantages and how can these be built on?
- What are the problems and how can these be reduced or overcome?
- What information do you need to prepare for visiting industrialists?
- How can you make effective use of industrial mentors?
- How can you make contacts with people in industry?
- How can partnerships be made sustainable? i.e. more secure than contacts between individuals.

Students

- What do you see as the benefits of working closely with people from outside of school?
- What aspects of your course do you feel need support from outside the school?
- How can these people help you with this?
- What problems or difficulties do you foresee?
- Can these be overcome? How? How important is it to find a way around these problems? Who should find the solutions?

Session 2

Purposes:

- Finding mutual benefits for schools, industry and students.
- Sharing involvement in course planning.
- Being realistic about what is achievable.

Some objectives for school/college – industry links:

- To achieve, at a local level, better personal understanding between people in education and industry.
- To develop a better understanding of each other's needs.
- Working with a team who contribute to students' learning in school.
- To use people from industry to make course delivery more effective and to meet further objectives of the course(s).
- To use people from industry to improve the range and quality of students' work.
- To assist with the professional development of people in education and industry.
- To improve the employability of students.

Tasks

Working with a team who contribute to students' learning in and out of school:

- Discuss which of these objectives are appropriate.
- Are there others you want to add?
- Put your objectives in an order of priority.
- Attach a time scale for meeting these objectives.
- Use this to develop an Action Plan:
  - what will be done?
  - who will do it?
  - when will it be done?
  - what will the outcomes be?
  - how will this benefit students, teachers, people from industry?

Suggestions

- Make use of personal and group visits.
- Use an evaluation to plan for improvements for the future.
- Look for unexpected benefits as well as those expected.
- Develop early in the initiative a way of monitoring/measuring how well the benefits/objectives have been met.
- Make use of feedback from all concerned.

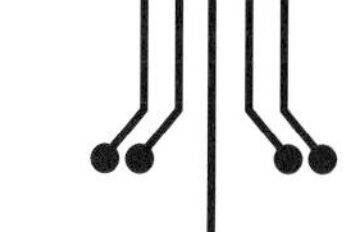

Case Studies: Education/industry collaboration in the RCA Project

See also the two case studies in the Advanced Manufacturing, Design and Technology student book; Working with industry – students describe their work with Asda Supermarkets and PDSA (pages 220–226).

CASE STUDY 1 – Working with the food industry

In the development of the RCA Schools Technology Project, the project team and teacher fellows developed a number of links with industry. This included visiting companies and factories, experiencing teacher placements, meeting product designers and product managers, discussing processes, following a product being developed, collecting information and writing together. The reciprocal arrangements for partners in industry to visit schools and become involved in the education of the next generation meant that both parties had a much clearer understanding of the other's work.

This experience for our teacher fellows meant they:

- improved their own teaching and were able to write focused practical tasks and designing and making assignments which were more accurate and realistic
- gained access to resources for their lessons and for the Project's publications which were not readily available
- updated their professional expertise and changed their viewpoint including stereotypical images of industry and common misconceptions
- were able to use the company's developed product and standards to provide a framework for student work
- gained confidence in the classroom to address these new areas as they were able to talk to students about what they had seen and experienced.
- developed a shared understanding of manufacturing when they carried out visits with teachers of other specialisms (e.g. food and systems and control) which led to many common starting points and joint projects.

Introducing industry to students

The following mechanisms can be used to introduce students to industry in any scheme of work and feature in the RCA course:

1. Designing and making assignments (DMAs) in an industrial context.
2. Case studies from industry of product development and manufacturing.
3. Product evaluation activities using examples from industry.
4. Introducing industrial-standard equipment.
5. Visits and outside speakers.

1 DMAs in an industrial context

A partnership with Pennine Foods resulted in a rich DMA where students are asked to design and make a new cook-chill pasta dish as a prototype for manufacture. The unit of work features focused practical tasks and a case study written in collaboration with Pennine Foods. Our work with them highlighted to us the importance of students becoming more exact about the type of materials, the quantities, the processes used and the controls that come into operation during production, as well as cost implications. They provided us with much needed information about production planning, HACCP, flow diagrams, process control, scaling-up and so on.

A plan of the factory layout was useful, so that students were able to understand a complete food production system in operation, from the intake of raw materials to the output of the finished product. Within this system sub-systems were identified. For example, where parts of the product are manufactured, and then combined on an assembly line, e.g. pasta, plus sauce, plus flavouring. Within each sub-system there were various controls that must operate and these will be part of the quality plan or safety plan for the product. Students can be asked to set up a production line simulation.

Our team included other specialist teachers. A systems and control specialist developed a task to mirror the weight control mechanism which checks the weight of all the made-up pasta dishes on the production line. A resistant materials specialist devised a way for students to make a ravioli board which was seen in the test kitchen and used to test ravioli shapes and fill them more accurately.

2 Case studies from industry of product development and manufacturing

Case studies are essentially photographs, slides or video footage, with accompanying text about a specific company or product. They are extremely valuable because they open up the adult world for students and can be used in a number of ways by teachers. For example:

- As a basis for discussion to show how their designing and making mirrors the adult world.

⇨

- To build an awareness of differences between domestic, catering and volume production. To highlight the difference between designing and making for yourself and prototyping.
- As a visual (photostory) life story of a product, understanding the scale of production.
- To pose further questions (e.g. research for homework) or activities (e.g. students go to supermarket to find out about existing products).
- To explain difficult and otherwise abstract concepts (critical control points, sensory analysis, selection of materials according to properties, effect of temperature, food science, GANNT charts, JIT production).
- To give examples of real specifications, mood boards, costings, market research, marketing, quality assurance schemes, flowcharts.
- To simulate production/assembly lines.
- To set targets for able students (getting it closer to industry specification).
- To look at materials – compare fresh, powders, liquids (e.g. flavourings).
- To use prior to a visit or a talk by a company.
- To help with evaluating existing products.
- To introduce industry standard equipment (multi moulds, weigh machines, conveyor ovens, mixing machines, pasteuriser), terminology and procedures (e.g. sensory analysis, HACCP).

3 Product evaluation activities – industrial examples

Looking at existing products is a valuable way to explore how students think things might be made. They can use their experience of products to add to their skills and knowledge and draw on this in their own designing and making. Many product designers use a test panel of all the competitors' products as a starting point for their designing.

Information about industrial examples can be used with students to:

- identify functional parts – why is this product like this?
- understand the demands which were placed upon the product developer
- look at the criteria used to judge the products and how improvements were made
- understand how products change over time and reasons for the changes
- give them ideas to draw upon in their own designing and making.

Industrial examples of where products have come from are also interesting for students – Greenhalgh's bakery looked at their waste puff pastry from the manufacture of their pies (as this was costing a lot of money) and developed a new product to use it up. Pennine foods developed a gaucamole to use the avocados that were too ripe for prawn and avocado sandwiches.

4 Introducing industrial standard equipment

In addition to securing information to support the Project's materials, we felt it was important for teachers to have the right equipment in their classes, and have worked with a number of suppliers to help develop their products in an appropriate way for the classroom, and support them with projects in our books which make effective use of the equipment. We have also encouraged suppliers to provide training and support (e.g. a help desk) in order for teachers to feel competent and confident when using equipment.

This has included specialist companies who supply monitoring equipment, e.g. pH, HACCP software and training resources, systems and control, specialist bakeware, measuring equipment.

For example, Dohler supply specialist ingredients for the food industry, such as flavourings. Armfield already supply training equipment for higher education and industry and have developed a number of scaled down processing units which will be invaluable for schools (such as a miniature pasteuriser). With this equipment schools really can model manufacturing processes.

5 Visits and outside speakers

Students' greatest benefit would come from visits to factories manufacturing food products. Time constraints can make this difficult and a friendly factory may not stay friendly and welcoming for long if inundated with hundreds of students from local schools. An alternative solution is for a teacher to go and transmit this information to the students, or for a company to come and talk to students.

Teacher visits

Take photographs in industry to display, enlarge and use with students to point out how a product is made. The photos can be used to show the manufacturing process in order. Photos can be enlarged and highlighted to make a particular point, e.g. a photo showing people assembling a pasta dish, one ⇨

CASE STUDY 1 – *continued*

showing the thermometers used to check the temperature of the soup, look at how many people are used to put this product together etc. Some companies have training or promotional videos which may be useful.

Company speakers

This works best where the speaker has been properly briefed. If possible let them come to a lesson beforehand, if not show the speaker examples of the students' work so that they are able to judge the level and discuss with them how to pitch their talk. Encourage them to bring slides/videos and samples and other visual aids. Many speakers have children of their own and are able to relate well to students when well prepared.

Student and teacher visits

Visits must be well planned and a visit beforehand by the teacher is recommended. In this way a teacher can look at the product in detail and understand the different stages – machinery used, the control systems, temperature sensors and control. Discussion beforehand with the students is essential – what they will see, what to look out for. A worksheet asking students questions relevant to the visit, picking out particular things like how a certain machine forms the containers, how and why any foods are stored in the warehouse and how the product is date stamped – is essential.

Take another D&T teacher with a different specialism with you. Two areas can be covered in one visit, e.g. the food person can extend their knowledge of control mechanisms and pneumatic systems. Relate to work going on in other areas of the D&T curriculum: if students have made something using a control system, bring that into the conversation when trying to explain about the control systems they will see on the visit.

CASE STUDY 2 – Working with the textiles industry

The Textile Industry in Partnership with Schools (TIPS) Project is a joint initiative of the North Nottinghamshire TEC and Nottinghamshire and Derbyshire Clothing and Textile Association (NADCAT). Its aims are:

- to enhance links between staff and students in schools with local textile employers
- to promote a wider understanding of the range of career options available in the textiles industry locally
- to provide professional development for teaching staff through a structured training programme and provision of comprehensive curriculum support materials
- to raise students' aspirations and achievement.

The project works by linking each school with a local textile company to form a partnership. Students, with support from their teachers, develop a mini-company and, working to a brief from the employer, undertake market research, shop reports, design ideas, worksheets and garment ideas. Each group will develop work skills and an understanding of how the textile industry operates. This work can be carried out at GCSE or Advanced level.

The Project requires a commitment from:

- the relevant teacher from each participating school to attend the Project Planning Day and to undertake a teacher placement of at least 3 days with the business partner
- the school to provide a minimum of five students participating students to visit the partner on at least one occasion during the project and to undertake a minimum of one day (ideally one week) work experience/work shadowing with the partner business (or other suitable business during the project)
- students and teachers to contribute to the evaluation
- schools and business to plan the project together.

The business is asked to:

- provide a named contact person to liaise with the partner school
- facilitate the teacher placement and student work experience/work shadowing
- visit the partner school at least three times during the project
- assist the partner school with sourcing – materials and expertise
- to give consideration to participation in the local EBP and participation in local mentoring initiatives.

The Project also provides support from expert business advisers and education specialists.

Each of the projects operated over a period of about 8 months. One of the keys to the success of the projects is the requirement to develop and keep to a set of milestones. These are dates when key things happen; they are agreed at the beginning of the project and everyone knows what needs to be done, by whom and by when.

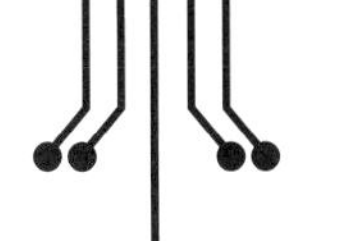

CASE STUDY 3 – Hewlett Packard Computer Peripherals, Bristol

Hewlett Packard decided that they wanted to work closely with the group of schools in the local area. This included seven secondary schools as well as all primary schools in the area. Because of the nature of the company, much of the work was IT based but the company also recognised the importance of D&T and some of the difficulties of delivering the course at Advanced level. They were also very keen on promoting what they saw as essential skills such as teamwork, problem-solving and decision-making skills.

They contacted all seven schools to invite them to take part in a one day activity for all Advanced D&T students in the area.

The activity was very carefully planned involving a series of after–school meetings over several months attended by engineers from the company, teachers from all of the schools and a representative of the local education authority with responsibility for curriculum development in science and technology.

The day that was planned for students involved:

- a series of team building activities – each student worked in a team with students from other schools
- an activity in parallel problem solving – to identify two faults in a hard-disk drive assembly that the company had only recently solved themselves due to one problem masking the other
- presentations by each team focusing on how they identified the problem.

Each team was supported by an engineer from the company who acted as a facilitator.

The evaluation of the activity highlighted:

- the importance of good planning involving people from the schools and the company
- the motivation of the students arising from working on a 'real problem' and being treated like an adult by the company
- the benefits to the engineers from the company in developing their inter-personal skills.

CASE STUDY 4 – Working with primary schools

This activity was based on an exercise carried out by British Aerospace in Hertfordshire. It was intended to improve students' understanding of some of the issues involved with manufacturing components on different sites and bringing them together for assembly.

A group of Advanced students was asked to design a product containing a number of components that would be contracted to students in local primary schools for production. The components would then be brought to the secondary school for assembly. Each Advanced student worked with a group of students in a number of primary schools.

The product used was a small model aircraft that could be made from Balsa wood.

The Advanced students had to:

- find out the capability of the sub-contractor (the primary students)
- design the product including sub-assemblies
- produce a specification and drawings that were clear and could be used by the sub-contractor to produce each component to the correct standard and tolerance
- make sure that all of the different components would fit together
- learn how to work with a group of young students and manage them to assemble the final product.

Lessons learned

Some common strands can be extracted from these case studies leading to a set of success criteria that will help to make these partnerships sustainable:

- the importance of having a shared vision and goals – establishing the benefits to all involved and a commitment from all concerned
- the importance of good planning, and involving everyone in this
- the importance of effective and regular communications
- the importance of a project leader – someone taking responsibility for the management of the partnership
- the importance of setting milestones – key dates with what will be delivered and by whom; keeping to these milestones
- the importance of establishing clear evaluation criteria that can be easily monitored at the outset and involving everyone in the evaluation – the basic rule must be 'how can we do it better next time?'.

CHAPTER 7

USING THE STUDENT BOOK

The *Advanced Manufacturing Design and Technology* student book has five main components:

Chapter 1 Planning and managing your own learning
Chapter 2 Designing and manufacturing
Chapter 3 Designing
Chapter 4 Manufacturing
Chapter 5 Extended case studies

These are meant to work together rather than being treated independently. Throughout the book you will find some common features:

- Short case studies — used to illustrate key points in the text
- Focused tasks — used to help students 'make sense' of the text
- Do the right thing — activities to help students apply what they have learnt to their own work
- Stop and think — raising wider issues for students to think about related to the topic being covered.

Chapter 1
Planning, organising, managing and evaluating your learning

Equips students with the skills to be successful in their course as well as enhancing key skills. Develops student autonomy and relieves the pressure on the teacher

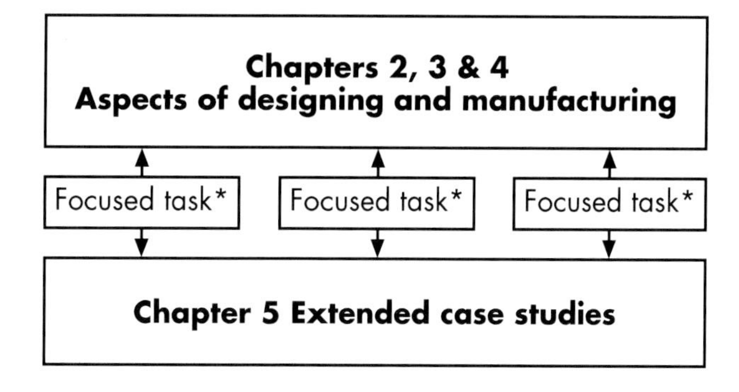

*FTs link the case studies to designing and manufacturing

This diagram shows some of the links between the different parts of the Student book

Using Chapter 1 – Planning, organising, managing and evaluating your own learning

This chapter can be used by students in several ways.

1 As a self-contained section used at the beginning of the course to provide them with basic skills in managing their work. This approach is unlikely to be totally successful as it is de-contextualised and has no immediate bearing on the rest of their work. It may seem to have the advantage of preparing students for what is to follow but the learning is likely to be superficial.
2 Using parts of the chapter to heighten students' awareness of the skills they need; providing them with a short focused assignment to provide a context for some of the activities, for example, developing an Action Plan for the assignment. As students are given further tasks and assignments particular aspects of Chapter 1 can be used to support the development of these organisational, management and other skills.
3 As a 'dip in' resource used by students on a 'need to know' basis. This is also likely to lead to a superficial level of understanding as the learning will be ad hoc, unstructured and have no progression.

It is strongly suggested is that approach 2 should be used extensively towards the beginning of the course, leading to approach 3 as students gain confidence and capability in organising and managing their own learning and work.

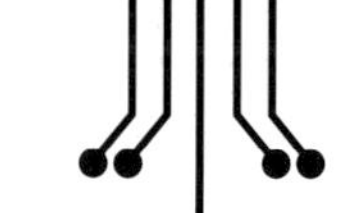

InSET – Planning the use of Chapter 1 of the student book

Read the chapter yourself to become familiar with its contents.

Alternatively, allocate parts of the chapter to different members of the course team. Ask each person to summarise and explain the key features of their section and its relevance to the course.

Using the plan and the summaries above, identify the parts that students need to be familiar with early in the course (key elements) and those that can be developed later.

Develop focused activities, relevant to the course, to provide the context for the key elements.

Map the other aspects of Chapter 1 to tasks and assignments that students will work on later in the course.

Plan how students will develop their skills and expertise in these areas.

Using Chapters 2–4 – Designing and making

A book of this nature cannot possibly hope to cover all aspects of designing and manufacturing appropriate for all Advanced courses. It therefore provides insights using examples of a wide range of products to establish generic principles. It can be used to support students' work on any Advanced course in Manufacturing, Design and Technology.

The approach taken is to work from a systems level to a product level.

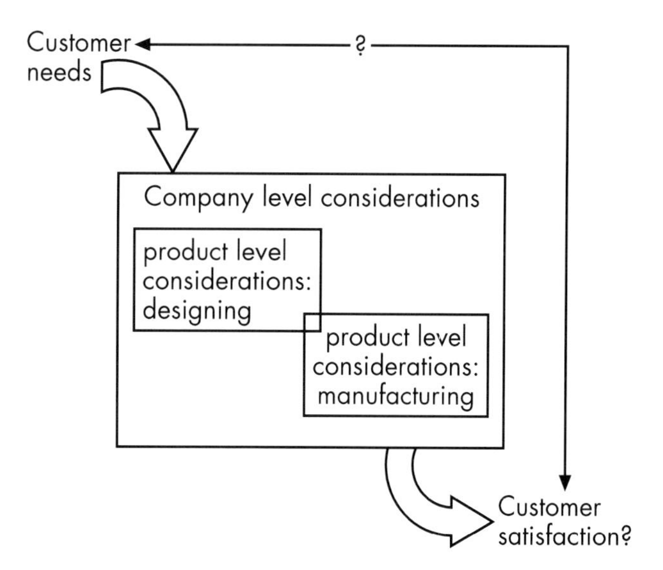

Manufacturing as a system – from customer need to customer satisfaction through manufactured products

Using Chapter 5 – Case studies

These are more extended case studies that often tell an interesting story to highlight key features of design and manufacturing; they are referred to at a number of different points throughout Chapter 2.

You can also use these case studies to highlight or illustrate particular points you want to make on your course.

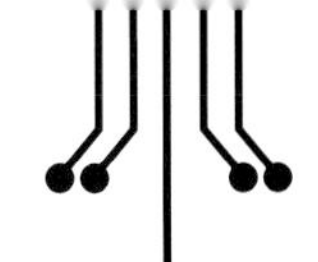

Linking chapters together

You will also find Focused Tasks in Chapter 4 linked to the more extended case studies in Chapter 5. These tasks ask students to look at one or more case studies, for example:

- to identify common features such as similarities in the designing approaches used across a range of different types of product
- to make comparisons across different volumes of production and different types of materials.

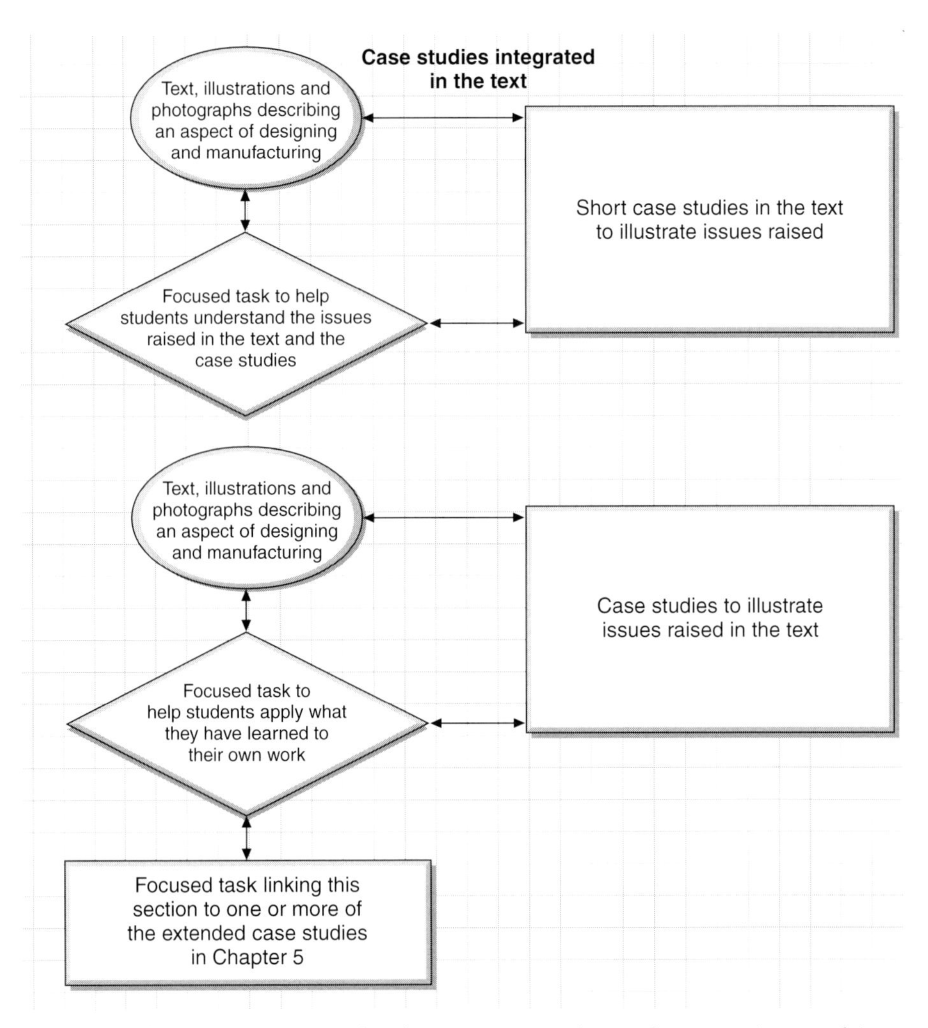

The layout of a typical section within the Designing and Manufacturing chapter of the Student book

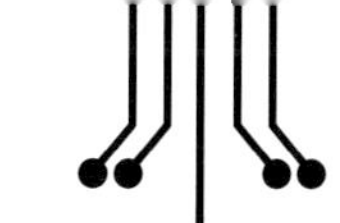

InSET – Planning the most effective use of the student book

Work as a teaching team involving all those participating, including industrial mentors.

Everyone should have their own copy of the RCA STP Advanced student book.

Everyone should be given sufficient time before the meeting to become familiar with the contents of the book. You may also find it useful to supply everyone with copies of the diagrams above giving an overview of the book and how the different parts relate to each other.

Develop a list of the key aspects of the course you are following:

- key areas of knowledge
- key aspects of designing and manufacturing
- key aspects of assessment, for example, major design and manufacture assignment, industrial case study.

Match each of these to the contents of the student book.

Use this to reach agreement about the best way to use the book.

You may need to:

- write your own focused tasks to draw out features appropriate to your course
- make reference to other resources available to the students
- develop your own short case studies based on local contacts.

CPD activity: Improving your own knowledge and understanding

The Project had access to a wide range of expertise, and many aspects of the student book describe areas of designing and manufacturing that may have developed since you acquired your own information about these. You can use the student book to update your own knowledge and understanding. This will be most effective where you use the book to establish the areas you want to find out more about and to help frame the questions you need to ask. You can extend your knowledge by working with your own industrial partners and contacts, through arranging further visits and through contacts with local Universities. At a simpler level, searching the World Wide Web through a search engine using keywords such as 'Manufacturing' can be very fruitful.